ECHOES

OF

FREEDOM

Best Regards,
Wendell Williams
Sept. 6, 2011

ECHOES OF FREEDOM

CLENDEL WILLIAMS

INTRODUCTION
ELAINE WILLIAMS CROCKETT

These creative narrative history stories, some humorous along with numerous facts, relate to the times and living conditions that were present during WWII, from 1942 through 1945.

BUILDERS OF LSTs 1942 – 1945
UNITED STATES NAVY SHIPYARD
EVANSVILLE, INDIANA

ECHOES OF FREEDOM

As Mark Twain stated in his writings, part of every story is based on an actual situation. The stories are written to instruct, educate, inspire, linking the past with the present.

Navy photos:
Evansville Museum and Arts, History and Science.
Wesner Studio:
Back cover photo of author.
Bryan Isenberg photos:
Interior photos of LST-325 Ship Memorial Inc.;
Boarding house and author, where he roomed 2 years.
Gantry crane, reminder of the WWII Navy Shipyard.
Vehicles photos from authors' private collection.

Several articles by the author, have appeared in the Bedford-Times-Mail; The Golden Age by Hoosier Uplands; The Hoosier Heritage Magazine; The Reader's Digest and Caves North of Campbellsburg. Other stories appear in the author's book, *BLUE ECHO MEMORIES*, published in 2005.

FIRST EDITION
ISBN 978-0-692-00609-2
Library of Congress Catalog Card Data:2011912248

Printed in the United States by Morris Publishing®
3212 East Highway 30
Kearney, NE 68847
1-800-650-7888

Swinging LST Bow into Position

The destinies of two great empires seemed to be tied up in some....things called LSTs.

Winston Churchill, Prime Minister of Great Britain
During World War II

An LST isn't such a glorious ship to look at. It's neither sleek nor fast not impressively big--no bigger that an ocean freighter--and yet it's a good ship and the crews aboard LSTs are proud of them.

Ernie Pyle, World War II Correspondent

Dedicated in Memory

of the men who fought on D-Day, to them we owe our freedom. When freedom had to be fought for or abandoned, they fought.

And to my
Miss America 1944

Edna Satkamp Williams
1921 – 2007
We never lose the ones we love
for even tho' they're gone.
Their memory, in the hearts that care
will ever linger on.

INTRODUCTION

Elaine Williams Crockett

Our way of life during the 1940's has rapidly slipped away from us. Coal stoves have been replaced by gas and electric central heating. Feather beds and straw ticks have been replaced with foam mattresses, and the washboard replaced by the automatic washer. But our heritage encompasses many things, not only our past way of life, but also our values, our traditions, and our history, and that's what we must preserve for future generations.

Thankfully we have this book that lets us peer through the window of time back to World War II with immediacy not often found in American history. In these pages, we not only experience the difficulties incurred by the men and women who served on our home front but we also experience some of the fun and spontaneity of their youth and the characters they encountered.

There's Fatso, the cigar-chomping, fun loving married man who decided to help the war effort by shacking up with a beautiful redhead to economize on the rent. Ma Burger, a motherly figure to the young men in her boarding house, also had a little black market business going on the side. And in the back woods of Kentucky, a woman who was challenged with the sanitation of keeping a pigsty near a baby replies indignantly, "We haven't lost a pig yet."

These are just some of the characters that bring these stories to life, along with several other faces of the many men and women who participated in the war effort. This portrayal of the people who fought for our country, with all their wonderful strengths and frailties, is exactly what makes this book so special.

Contents

Acknowledgements

The author gratefully acknowledges and extends his appreciation for the invaluable assistance of Stanley C. Coy, editor of Hoosier Heritage Magazine; Jane Clark, author and former publisher; Elaine W. Crockett, Attorney, United States Federal Trade Commission, Retired, and former WWII Marine Veteran and Educator Harry Baker, for proofreading and correcting errors in grammar and punctuation.

I thank Marji Morris, retired teacher and Director of Washington County SOAR and also of the Writer's Bloc Group whose members assisted with their suggestions. Thanks to Shelley Crockett, former Director of Internet Marketing for Warren Bros. Movie Studios, for the book title suggestion.

Recognition and thanks goes to the Indiana Magazine of History for permission to use excerpts from the December 2008 article "The Most Wonderful Thing I Have Ever Seen: Indiana's Contributions to Petrified Man Hoaxes," by Carl Runyon and Randy K. Mills.

Thanks to Thomas R. Lonnberg, Curator of History, Evansville Museum of Arts, History and Science for providing the Navy pictures of Landing Ship Tanks during construction at the Evansville Shipyards and for the war zone pictures of LSTs making beach landings.

I appreciated the extremely helpful volunteer LST veterans on my three guided tours of the LST Memorial Ship 325 located at 1801 Waterworks

Road, Evansville, Indiana. Walking the decks and climbing nearly vertical stairs brought back memories of my employment during the construction of the 167 LSTs built at the Evansville Shipyards that earned 309 Battle Stars during WW II, 27 Battle Stars during the Korean War, and 35 Battle Stars during the Vietnam War.

As a member of the United States LST Association, I receive the LST Scuttlebutt Periodical publication that is hard to put down, as the articles are historically fascinating.

Others who provided encouragement and assistance during my three years of coordinating these stories are: Genevieve Lynn, Ernie and Mary Morris, my son Ryan David Williams, and my daughters, Nancy Kay Isenberg and Marla Jane Drane for their knowledge of computers, and to Jon Morris, for coming to my rescue when the devil got in the computer locking it up.

I appreciate the assistance of Charlotte Roesner, Curator of the Mesker Park Zoo and Botanic Garden. Also, thanks to researcher Barb Manzi, for providing verification of several events.

Thanks to Mike Purkhiser and Lowell Young, for their individual contribution to several stories, along with Jack and Kathryn Sullivan, for their assistance and encouragement. Thanks to Kaitlyn Wesner, of Wesner Photo Studio, for taking my picture that appears on the back cover.

Finally, I appreciate the Morris Publishing Guide and their consultants as they make self-publishing the answer to being a published author!

Prologue
What is an LST?

LANDING SHIP TANKS WERE FLAT BOTTOMED AMPHIBIOUS BOATS.

The boats had no official names, just numbers, such as LST 325. Those of you who have seen a barge being pushed or towed on the Ohio or Mississippi Rivers have seen the type of construction that comprised the concept of a flat-bottomed LST.

The LST was the first of its kind. The original idea was the dream of Winston Churchill during World War I. He had the coal carrier, The River Clyde, converted into a war vessel with large doors in the bow. It was a failure as the bow keel design was of the 'V' type that could not beach in shallow water. Most of the 2,000 British troops drowned in 1915 as they jumped with full gear into deepwater at the Gallipolis Peninsula in Turkey. Nonetheless, the failure of Churchill's idea confirmed the usefulness of a flat bottom type of ship that could be beached.

In the fall of 1941, as Hitler escalated the war against Britain with buzz bombs, the British Navy and the United States Navy pooled ideas on how to develop such an amphibious vessel. The basic barge design proved to be the best plan for the 1,051 LSTs built during WWII.

It was actually a simple ship. A top military secret was not necessary in the design of the ship, but rather in the proposed method of operating the

transfer of large quantities of liquid ballast necessary to keep the ship afloat in all kinds of weather. The LST's basic purpose was to make beach head landings delivering army tanks, trucks, jeeps, and ammunition along with troops.

The dimensions of the LST were 327.9 feet in length, and 50 feet in width, weighing in at 1,625 tons. An LST could carry 2,100 tons of cargo, 13 General Sherman Tanks on the lower deck, and a total of 40 trucks and jeeps anchored and tarpped down on the top deck.

Two 900-horse, 12-cylinder locomotive type GM diesel engines, side by side with separate drive shafts and six-foot propeller blades, could move the 3,725 ton, fully loaded vessel through normal seas at 9 to 10 knots per hour. Top speed was listed at 11 knots, however, 8.5 knots was the average speed.

Although every LST was built according to the same basic design, various improvements were incorporated as the shipbuilding program progress-ed. The average cost to build a LST was $1,300,000.

The normal armament equipment on a LST was: One 3 inch (765mm) gun
Six 40 mm guns
Six 20 mm guns
Two 50 cal (12.7mm) machine guns
Four 30 cal (7.62mm) machine guns

To fully operate the craft around the clock it required eight to ten officers. Also needed were 100 to 115 sailors, most of whom were teenagers, to swab the decks, operate the engines, generators, pumps, prepare meals, stand watch and to man the guns.

One 'salty' laughed about his first ride on a LST while crossing the Atlantic to North Africa during the war. He related, "An LST had a blunt bow and a shallow draft, and when she headed into heavy waves it caused her to lose headway momentarily and to shimmy and vibrate as she was stunned by super large waves. At times the 328 feet of steel actually twisted when bucking a gigantic wave, depending on how heavy the ship was loaded."

He continued, "Once I drew the four-hour watch in the least desirable spot, the crow's nest, located high up on the mast in a barrel high above the officer's quarters. We encountered a severe tropical depression with visibility nearly zero. The ship seemed to be part submarine as it plowed head on into 30 to 40 foot waves accompanied by winds of 45 to 65 knots. Salt spray pelted my face, as I tried to keep a sharp eye out for planes and submarines. Waves with a thunderous roar broke across the main deck dozens of times. No roller coaster holds any comparison to my frightening ride tossed in the crow's nest for several hours, and the ride was free. The dry heaves had me considering the thought that dying would be a great relief, and I had no bucket up there in the crows nest as I lost my appetite."

While making a tour of LST 325, one the guides told of his experiences. "At times, high seas caused the ship propellers to spin out of the water with a screaming whine, as the engines speeded up until the governors shut them down. Then the diesel mechanics had to reset the engines' governors and get the ship back up to normal operating speed."

According to what I have read in the LST Scuttlebutt Publication, keeping the ship on course during a tropical storm proved to be a difficult task. In order to quickly change the direction of the 328-foot ship, it was necessary for the diesel mechanics to assist the steering by running one engine in reverse while the other engine was in forward gear. With their assistance, a quick turn to the right or left could be navigated.

An old friend asked me recently what I was doing to pass away the time and when I told him that I was writing stories about LST boats, he nearly went berserk. "I have no desire to ever see one of those ships again. I served on several different LSTs as a diesel mechanic. They called us the 'black gang' as after four hours in the engine room, we were covered with grease and oil. I was never so sick in all my whole life as I was riding those darn tubs.

An estimated 230 LSTs participated at the Normandy D-day landings. The amphibious vehicles with their gallant crews won the "Victory at Sea" in the South Pacific. During the war, the LSTs furthered the cause of peace in many different ways. Several were converted to hospital ships.

Of the 1051 LSTs built, 670 were built at five inland yards located at Ambridge, PA; Pittsburgh, PA; Seneca, IL and Jeffersonville, IN where 123 LSTs were built, while Evansville built 167 LSTs.

This book relates several unique stories about the men and women who served our country at the Evansville Shipyards during WWII. A nickname for the letters LST was "Large Slow Target."

Author's Foreword

These creative narrative history stories, some humorous along with numerous facts, relate to the times and conditions that prevailed during WWII from 1942 through 1945. They are a part of our American heritage.

An attempt has been made to provide an understanding of the working and living conditions that prevailed during those years.

The federal government instigated daylight saving time the year round, so there was no confusion of changing the clocks, winter and summer.

Several chapters reveal the true human character, endeavor, circumstances and conduct of the people who made up the "invasion work force" on the home front. Most everyone can visualize their parents, grandparents, aunts, and uncles in the general society to which they belonged during that period of history. The people had been hardened by the hardships of the depression during the thirties but they still had faith and determination,

As Paul Harvey might say; "And now, for the rest of the story."

On Sunday, December 7, 1941, Japan dropped bombs on Pearl Harbor, Hawaii, and it changed the world forever. When President Roosevelt declared war on the Axis, the countries aligned against the United Nations, the American people unified and rallied in the war effort.

Everyone felt it was their patriotic duty as Americans on the home front to work together and to win that war. They donated to the scrap iron drive, the rubber drive with old tires and girdles. Housewives turned in their scrap aluminum pots and pans and were asked to save fats and grease for the war effort.

Throughout the nation, county governments donated valuable relics from the Civil War. Many grade school children, pulling their little red wagons, collected items from their parents, grandparents, and neighbors as their patriotic duty.

Everyone was urged to buy government war bonds in order to finance the war. To buy them one became a true soldier of Democracy.

The government imposed rationing of all items and material that were needed in the war effort. New vehicles and rubber products including shoes were immediately rationed. On March 29, 1942, meat, butter, and cheese were rationed. Gasoline was rationed on July 22, 1942.

A new type of amphibious landing craft was to be built. This vessel was the LST (Landing Ship Tanks) that earned its proper share of honors during WWII.

In January of 1942, the United States Navy selected Evansville, Indiana as one of the sites for an inland shipyard. Evansville was chosen because it offered a favorable location on an inland waterway and had resources of skilled labor and established industries for subcontracting.

According to a lifelong resident, part of the site selected for the shipyard was formally a city dump.

The Missouri Valley Bridge & Iron Co. Leavenworth, Kansas was awarded the contract to build the LST in Evansville, Indiana. Temporary offices were opened on Saturday, February 14, 1942 at the McCurdy Hotel.

The first major installation was a bank of transformers. Floodlights were necessary for construction to continue around the clock, seven days a week. Surveying was begun. Thousands of yards of earth were trucked in. Grading of the ground continued as a pile driver sank oak timbers, 14 to 16 feet in length, firming up the ground in a few areas.

Concrete construction workers tied together thousands of steel reinforcement rods. An estimated 200,000 yards of concrete were poured to reinforce the twelve-acre construction site on which the boats were to be built.

The shipyard workers were actually the makers of history, whether or not they knew anything other than their respective daily job obligations. During the first few months of the construction of the LST boats, most of the workers groped around in confusion, not having any idea what the finished product would look like.

Work on the cofferdam was progressing and the fourth ship keel, LST-160 was laid on the day of my employment, Tuesday, July 21, 1942.

I worked on the swing shift for a year and then one year on days. The third year I worked the graveyard shift with the Skegg welding crew. There were times at night when the shipyards were encased in a gray mist so thick that the only boat visible was the one you were on. It was an eerie feeling, locating the staircase at the end of the work shift, in order to get off the boat.

About that time, a foghorn sounded out on the river causing chills to run up my back tingling my scalp. Almost immediately a muffled answer was heard from a barge, moaning like a wounded bull, a mile or more downstream.

The constant flash of dozens of welders across the yards lit up the gray mist producing weird kaleidoscope patterns when actually nothing was there. The muffled banging of steel against steel and the rat-a-tat tat of the chippers' air hammers echoed from warehouse rooftops, producing a scene of mystery and intrigue.

Another fascinating sight while working the graveyard shift was on foggy mornings as thousands of workers clocked in for the day shift. They moved silently about, through the mist, to their appointed destinations. The pea-soup fog drifted with the breeze as it crept in from the river.

I was first hired as a weld checker and soon afterwards attended welding school for a few weeks.

After having passed my flat weld test, I was assigned work with the shipfitters. It was my job to tack weld the deck plates after they had pulled the pre-assembled sections together using steamboat ratchet jacks.

Later, having passed the overhead and vertical welding tests I was presented a certificate stating that I was a certified three-position welder.

Part of the time I was on the pickup welding

crew. We had to finish welding what other welders had left unfinished. Those welds were usually in the most difficult places.

Once in a while, I would get a decent welding assignment and enjoyed being outside during good weather. Running stainless steel flat welds on the gun turrets was an enjoyable job.

The one welding job that I always thought would be fantastic to get, but was never asked to do, was to weld the outline of the ship numbers on the port and starboard sides near the bow.

It was an amazing sight to observe the rapid changes in the building of the ships from one day to the next. The parts of a ship by themselves would sink but when welded together they formed a craft that floated.

The city of Evansville was much like any other city of its size in 1942. Drive-in-theaters and fast food restaurants had not arrived in Mid-America. There were no shopping malls, and television had not been perfected yet. Rayon and silk was no longer available as they were used in producing parachutes for the air force.

Women wore their old silk hose with runs, and bobby socks were in style. Many of the young ladies went barelegged, painting a seam down the back of their calf with an eyebrow pencil. A slight curve in the seam gave it more of a realistic look.

Women dressed as women, wearing clothing consisted of dresses, skirts, and blouses. Ladies in business suits were seldom seem. Pants were worn in the defense factories. Women employed in the offices wore dresses below their knees.

Stores remained open until 9 PM on Monday and Friday nights to accommodate the war workers. On those nights as well on weekends, the sidewalks were usually jammed with sailors waiting for their ship to be launched and commissioned.

On weekends, Soldiers from Camp Campbell south of Hopkinsville, Ky., and also those from Camp Breckinridge near Morganfield, Ky., added big numbers to the population of Evansville.

The Sabbath was a day set apart as the day of worship. Usually, Sunday mornings were peaceful and quiet as church bells tolled across the city. The chimes of the Catholic and Lutheran Churches were soothing to the ear, even as they mingled with the banging sounds of the shipyards.

Stores were closed on Sunday, other than the Jewish store located on Second Street just one block off Main Street. It was closed on Saturday, which was the Jewish day of worship.

For entertainment there were at least six movie theaters in the city. Franklin Bowling Lanes and Burdette Park on the west side of Evansville, along with the Penny Arcade, the Carousel, and Mesker Zoo were great attractions.

Hill's Restaurant at the foot of Main Street at Dress Plaza and Riverside Drive was famous for their frozen malts. Jo-Jo's, near the airport on 41, the Brown Derby south on U.S. 41 and the Starlite Tavern at the top of the hill on West St. Joseph Street were each famous for serving wonderful food.

One Saturday night, the stars of the Grand Ole Opry from Nashville's WSM performed on stage with Roy Acuff singing *The Great Speckled Bird,* along with Red Foley and Ernest Tubb.

Minnie Pearl seemed to be the center of attraction with her high-pitch voice and a price tag hanging from her wide-brimmed hat. The Duke of Paducah kept saying, "These shoes air'a kill'n me!"

The Saturday night wresting matches were well attended at the Coliseum. Professional wrestlers traveled a circuit from St. Louis, Indianapolis, Louisville, and Evansville. After a month, they would move on and another bunch would appear. The promoters knew how to get the crowd stirred up with sound effects. With the audience yelling and screaming, several men were ready to get in the ring and do some real slugging.

West Franklin Street, west of the Pigeon Creek Bridge was always a busy place. There were six blocks of stores including the Franklin Lanes, a movie theater, a dime store, the Ford Garage, and a large Kroger grocery store. Franklin Drugs was located on the corner of West Franklin and St. Joe.

Main Street was not to be left out for entertainment. There was a cafeteria just off Main on 3rd Street, and there were three movie theaters, as well as a Ten-Pin Alley located near 7th Street.

Window-shopping with your sweetie on a Sunday afternoon stroll along Main Street was safe as the stores were closed.

At times, two supposedly blind men vied for alms. One Saturday afternoon, I heard one man yell at the other blind man, "Get on the other side of the street! This is my corner!"

At least twice a year the traveling organ grinder, cranking his portable organ, along with his pet monkey dressed in an abbreviated bellboy uniform, would show up along Main Street. The trained monkey would do a few simple tricks in tune with the music. When the music stopped, the monkey, who was on a lease, would hold a tin cup hoping for a donation. When someone dropped a coin into his cup, the monkey would tip his little red hat as if to say thank you.

As a reminder, as you read these stories, you need to keep in mind that the United States was still in the dark ages in some respects as compared to today's electronic global satellite era.

A penny postcard helped to keep families and friends in touch. A three-cent stamp would send a letter next door or across the continental 48 states within a week. A six-cent stamp would send an airmail letter over-sea to a person in service.

During the war, people planted Victory vegetable gardens for additional food in unusual places such as between the front walk and the house, as the average factory worker's income was less than $25.00 a week.

The railroads were the nation's only mass transit carrying passengers and supplies across the entire country. Troop trains shuffled soldiers day and night across the nation.

Everyone felt the effects of the war and the fight was on for liberty and justice for all.

Chapter

1

Arriving at N.Y.A.Camp

ARRIVING IN RIVER CITY IN LATE AFTERNOON ON SUNDAY, JUNE 7, 1942, WAS AN ENTIRELY NEW EXPERIENCE FOR THIS COUNTRY BOY.

I was fresh out of high school and at the age of 19, it was only a matter of time before I would be receiving my greetings from Uncle Sam.

Not being financially able to attend college, I needed to get as much experience and education as possible to face the future. The National Youth Administration (N.Y.A.) Program guaranteed two meals, a cot and thirty dollars a month plus training.

The letter I received stated that I should report as soon as possible to the N.Y.A. Camp in Evansville, IN. I was to be met at the Greyhound Bus Station by a young fellow wearing a T-shirt with the letters N.Y.A on the front. He would drive me to the camp in an army khaki colored school bus.

I drove Dad's '38 Chevy with my parents to Hurley's Restaurant in Orleans, Indiana, where the Greyhound bus always made a stop. There I left my mother with misty eyes as I purchased a bus ticket to Evansville, Indiana.

With the ten dollars that my father had given me, I was on my own for the first time in my life.

My arrival at the Evansville Bus Depot was uneventful. I had no problem spotting the young man wearing a N.Y.A., T-shirt. We were about the same age. He said, "We need to stick around a few more minutes in case there are other fellows that might have arrived on one of the other buses." No one else showed up, so I was his only passenger.

Other than the khaki paint job, the camp bus appeared to be a Carpenter school bus body made in Mitchell, Indiana. It was very similar to the one that I had ridden in while attending high school. I was familiar with the rumbles and rattles of the bus including the stiff springs, making it a rough rider.

I was exuberant with anticipation at every intersection. As the camp bus headed out St. Joseph Street, I began to be concerned. It appeared that we were headed for the country. After passing a cemetery and going over the rise of a hill, I spotted the camp on the right hand side.

I had no idea that the N.Y.A Camp was located in the former Civilian Conservation Corp (C.C.C.) barracks across from the Mesker Park Zoo. I had visited the Zoo two years earlier on a tour bus with a group of high school students.

I met with the camp superintendent and was assigned cot number eleven in barrack number one. There were several other boys milling around and they introduced me to the cold-water shower and other vital necessary installations.

That night I wrote a postcard, "Mom and Dad, I'm here, but it ain't nothing like home. Love, Your Son."

Chapter

2

Hitch Hiking Home

HAVE YOU EVER BEEN DESPERATELY BROKE? I MEAN, STRANDED AMONG STRANGERS WITHOUT SUFFICIENT MEANS TO PAY FOR YOUR NEEDS? I HOPE NOT.

It's a feeling you will never forget. It will give you an uneasy feeling in your stomach, the same kind of bellyache as if you had eaten green apples, leaving a lump in your throat. I can sympathize with anyone in that situation.

I vividly remember the Saturday afternoon of June 27, 1942, when I made my way to the Evansville's Greyhound Bus Station.

I had less money in my pocket than the $3.40 needed for a bus ticket to Orleans, Indiana, where I wanted to go. I had been at the N.Y.A. camp three weeks and was homesick. I asked the ticket agent how far I could ride for $2.30. He said that a ticket to Haysville would be $2.18. That left me with twelve cents in my pocket.

As the bus pulled up at the little store building on the southeast corner of the crossroads in downtown Haysville, I had to get off.

A few seconds later the bus roared away, leaving me standing there nearly suffocated in dust and black diesel smoke. As the motor whines faded away, like a whipped pup I felt abandoned. Tears tumbled over my sun burnt cheeks. Blowing my nose, I said a prayer, and with renewed hope and misty eyes, took the first step on the long road towards home.

It was four-twenty in the afternoon. The temperature was in the upper seventies. A few high feathery clouds spotted the sky as I walked across State Road 56 at the east edge of Haysville near the Meyers Truck Body Shop.

Having only walked a few hundred feet, I heard a motor vehicle approaching from behind. Turning around quickly I held up my right arm thumbing for a ride.

A shiny black 1941 Oldsmobile sedan rolled to a stop. A Good Samaritan in a dark business suit, white shirt, and dark tie, reached over and unlocked the passenger door.

As I got in, he asked, "How far are you going?"

My quick reply was, "Paoli," not thinking that he might be going farther.

He answered, "I'm headed for Indianapolis."

Immediately, I realized that he would be going through Orleans and that would put me ten miles closer home. However, since I had said 'Paoli' I wasn't going to change my story. He might think that I was out for a joy ride and the truth wasn't in me. I was indeed thankful that he had stopped.

Before we arrived in French Lick, we had passed the bus that I had been riding on. Soon after five o'clock, we were in Paoli where I got out on the east side of the public square.

Walk I did, mile after mile after mile. I met two cars, however, none going in my direction. Near Millersburg an elderly farmer in a 1927 Chevrolet, which had been converted into an open-top farm truck, offered a ride that I gladly accepted. The motor sounded like an old John Deere tractor as it hit and missed, poking along at ten miles an hour. The evening sun was setting low in the western sky by the time we arrived at the feed mill in Livonia.

I headed north past the Livonia grade school building, as I trotted on towards Campbellsburg. I was almost home! Only thirteen more miles to go! The stars blinked in a clear blue sky and before long the moon provided a clear view of the road to follow. A soothing cool breeze helped to refresh my energy.

Dogs barked at nearly every house, but none ventured out onto the roadway. The Lord was with me as I prodded on. The town of Campbellsburg had rolled up its sidewalks. It was near midnight when I awakened my folks. I had walked at least twenty-two miles with nothing to eat nor drink.

Dad and Mom were glad to see me and after a big slice of mother's pumpkin pie, I slept like a log. I was up in time to attend Sunday school at Mt. Carmel Christian Church with my parents. Later that afternoon, they took me to Orleans to catch the Greyhound back to Evansville and the N.Y.A. Camp. Dad gave me bus fare plus a few dollars to tide me over until I received my once a month paycheck of thirty dollars from the camp.

Years later, I often thought back to my long walk home. After I was married and working as an auto body repairman, my wife, Edna and I had to pinch pennies in order to make ends meet. Several times I sold our car to pay the grocery bill. We always had suitable transportation, as I would purchase a cheaper car, fixing it up to resell later.

I heard my father complain so much about people borrowing money and never paying it back that I was determined to never ask him nor Edna's father for financial help. We wanted to make it on our own. It wasn't easy.

At times we followed my grandmother's method of makings ends meet. She said they borrowed from Peter to pay Paul and visa versa. On several occasions, I borrowed money on a car title from a finance company and then when it came due, we borrowed money from the bank to pay the finance company or visa versa. This was our normal routine in order to keep afloat financially.

Some twenty years later I was manager of a finance office and asked a local banker for a loan of three hundred dollars. He said that if I needed the money to come back the following day. I took him for his word and when I went back the next afternoon, he loaned me the money. He explained, that if I had been in trouble embezzling the finance company funds, he didn't intend to be an accessory to a crime. Like most good bankers, he was cagey.

By giving thanks for our many blessings, the Good Lord provided for our daily needs. There were days when we only had faith, hope, and love to keep us going. "Our happiness has been in the art of knowing how to do without what we couldn't afford.

Chapter

3

Applying For a Real Job

AFTER SIX WEEKS AT THE NATIONAL YOUTH ADMINISTRATION CAMP I COULD NOT SEE ANY LIGHT AT THE END OF THE TUNNEL.

I had heard by the "Camp Grapevine" that it would be no problem to get a job at the nearby, Evansville Navy shipyards. After lunch on Tuesday, July 21st, I walked the two miles into town.

I went to the Missouri Valley Bridge & Iron Co. employment office on West Franklin Street. The building had a sign hanging over the entrance; Werner Motor Sales, Studebaker Vehicles.

After standing in line for several minutes, I was handed an application to be filled out. I found a seat and attempted to answer a dozen questions.

The interview line was backed up nearly to the front door. Finally when I got to the front of the line, the interviewer only glanced at my application. There was no physical, nor an eye examination. One person took my picture and another one finger printed me. Within a few minutes, I was handed a badge with my picture and number 4214 on it. I looked like a convict, starry eyed with ruffled hair. I was instructed to report before 4:00 P.M. to Clock Alley #1 for a job assignment.

Did you ever see a blind dog hunting for a bone? Well I was like that dog. I knew about where the shipyards were located by the noise. However, I had no knowledge where Clock Alley #1 was located.

Time was of the essence. I had less than two hours to get back to camp, hunt up the Superintendent, and tell him why I would be late for bed check. He was like a father, kind, and considerate. He said I could stay and eat two meals each day until I received my first paycheck.

My only means of travel was 'shank's pony', so I ran most of the two and a half miles to the shipyards and fell in a long line with dozens of other new employees that said they too were to check in at Clock Alley # 1.

A guard checked off my badge number on his list of new hires, and handed me a yellow slip of paper with instructions for me to report to a Cliff Limeberry, welding foreman on Way # 3.

Way # 3? What was a Way? Where was it located? What did this Limeberry fellow look like? Was he tall, short, fat or thin? My mind was in a muddle. I needed to find an outhouse first before hunting for a total stranger on Way 3, whatever and wherever that was.

Upon asking, I was told that the word 'Way' in shipyard logo was the location upon which the ships were being constructed.

Foreman Limeberry was easy to spot, as he was the only man that fitted the description I had been given. I had been told to look for a fairly tall slender man wearing a Stephen Hat and matching tan gabardine pants and shirt with a gold fountain pen and four cigars in his shirt pocket.

After waiting a few minutes until he finished talking with another new employee, I introduced myself and handed him the slip of paper I had been given by the guard.

Mr. Limeberry welcomed me aboard and assigned me as his assistant with the title of weld checker. My hourly rate would be the minimum rate of 50-cents per hour. After deductions of $1.40 which I understood was for Old Age Benefits, (OAB), later referred to as Social Security) I would receive $18.60 for a five day 40-hour work week.

Now that I had a job, I'd need to start checking the newspaper advertisements to locate a room where I could sleep, other than at the N.Y.A. camp.

~~~~~~~TAKE A BREAK ~~~~~~~

Writer: I took up writing full-time a year ago.

Friend: Have you sold anything?

Writer: Yes----my TV, all of my furniture, and recently I had to sell my house.
~~~~~~~

The Missouri Valley Bridge and Iron Company, July 21, 1942 employed the Author.

Below is a replica of the official shipyard badge. The original badge was turned in the second week of July 1945 upon leaving.

Clendel Williams, age 19

BUILDERS OF LSTs 1942 -- 1945
EVANSVILLE, INDIANA

Chapter

4

Sent To Get a Sky Hook

ON MY FIRST DAY ON THE JOB, MY BOSS SENT ME TO THE OTHER END OF THE SHIPYARDS TO GET HIM A SKY HOOK.

I was anxious to please my new boss, so I followed his instructions eagerly, since this was my first assignment.

He told me, "Go down on Dock 3, and hunt up the foreman. Tell him that I need a skyhook up here on Way 3."

Well, I wasn't moving very fast as I continued to think over his request. The more I thought, the more I questioned his sincerity. I remembered some years back when several of the neighbor kids and I were at a party. We were given a paper sack and a flashlight and told to go snipe hunting. We couldn't come back until we found at least one snipe.

I didn't know what a snipe was. None of the other kids knew either, except one boy who was a year older and smarter than the rest of us. He said, "There aren't any snipes around here. A snipe is a water bird with a long pointed beak, and I have never seen one."

So the more I thought about it, the more I had the feeling that the foreman was giving me the run around. This must be his little practical joke, since this was my first day on the job.

About half way to the dock area, I met a friendly face. I felt he would level with me. This fellow was about my height but considerably older than I was. I smiled and waved my right hand as I said, "Hi! I beg your pardon, but tell me; is there such a thing as a skyhook?" He gave me a big smile and chuckled as he answered me with a question, "Is this your first day on the job?"

"Err...Ye...Yes, it is," I stammered.

"Did you ever go snipe hunting?" he inquired.

Hesitating, I answered, "Yes."

He laughed, "Well, it's about the same. Some one is having fun giving you the run around." He chuckled, "Tell him that they are on back order along with about a hundred other requests for skyhooks."

"Thanks, I'll do that." I walked on down to where they were driving steel piling for the dock.

After fooling around, looking everything over, and checking in at the outhouse down on the riverbank, I reported back after lunch break.

I hunted up my foreman and told him, "They said the skyhooks are on back order and as soon as they come in, you can get one." He smiled and nodded his head letting me know that he heard me.

That was my initiation to the work force at the shipyards, and as far as I know those skyhooks are still on back order.

Chapter

5

The Fish That Couldn't Get Away!

MY FIRST DUTY EACH DAY WAS TO REPORT TO THE WELDING FOREMAN.

As the foreman assigned each welder a work location, it was my job to write his or her name, and clock number on the clipboard form that he had provided. Within an hour I was to go around and check on each welder to make sure that they were working at the correct location.

Then an hour before the shift ended, I was to go around and locate each welder and inquire as to how many feet of welding they had completed. At the end of the work shift I turned my report in to the welding foreman.

During my spare time, I wandered around the shipyard's forty acres with my hard hat on without anyone inquiring as to what I was actually doing.

It was fascinating to see thousands of workers doing so many different jobs at the same time and the progress that was being made each day.

Each evening I kept hearing a loud clanking, banging, and hissing noise. At times the sound even surpassed the booming sledge hammers against hollow metal compartments, the rumble of the overhead gantry cranes, and the ear-piercing chatter of air hammers used by metal chippers.

One evening after lunch, I decided to find out what was making all of the loud noise. I discovered a tugboat operating a steam pile driver installing steel caissons in the Ohio River, a distance of three hundred feet. A cofferdam was being built.

After I had completed my routine rounds every evening, I would spend an hour or more watching the completion of the dam. Once it was finished, they started pumping the river water out where the launching ways were to be built.

With four large pumps operating around the clock, it took three days to pump out several million gallons of river water impounded by the cofferdam. All that was left were hundreds of fish splashing around in three feet of soupy mud. They reminded me of maggots that I had seen, squirming around in a bucket of rotten tomatoes.

A dragline scooped up hundreds of buckets loaded with live fish along with the mud, dumping them back into the river. Finally, after two days the only fish visible was a large one splashing around in three feet of muddy slush.

That evening, as the sun was setting low with its streaks of red and gold across the western skyline, I watched two men in a small boat attempt to subdue this huge water creature.

In its final efforts to survive, the fish's gigantic tail splashed mud high into the air. Both men, wearing hip boots and rain coats, with their hard hats strapped on under their chin, were covered with slimly mud from head to foot as they stood up in the boat.

It appeared that one man had two paddles stuck into the mud trying to stabilize the boat while the other man was holding a harpoon. They had the fish corralled into a corner of the muddy pond, created by the cofferdam. Finally, they managed to stab the fish with the harpoon,

As daylight ebbed, the shipyard's floodlights came on creating ghostly shadows where none previously had existed. The two fishermen appeared as animated silhouettes, holding on to each other against the ribs of the cofferdam.

It was time that I got back to work!

The next afternoon, as I was clocking in, everyone was talking about the large catfish that had been caught the night before. They said that it weighed sixty-eight pounds,

Wow! What a fish story about a fish that didn't have one chance in a million to get away!

~~~~~~~TAKE A BREAK ~~~~~~~

Writer: Do you know it took me over
twenty years to find out I
actually had no writing ability?

Friend: So what did you do---give it up?

Writer: Oh, no, by then I was so famous
I couldn't afford to.
~~~~~~~

Chapter

6

Mesker Park Scare

FLASHES OF LIGHTNING ZIGZAGGED ACROSS THE SKY WITH CONTINOUS BOOMING OF THUNDER VIBRATING THE EARTH WHERE I WAS STANDING.

At first a few raindrops fell. Suddenly, the heavens opened up, drenching my entire body. The swirling rain and wind blinded me as I walked.

It was the third full week of July in 1942. I had only worked a few nights as a weld checker on the swing shift at the Evansville Navy Shipyards.

Working half the night was a new experience for me. I was nineteen years old. This was my first real job, so come what may; I had to tough it out.

During the evening hours several inches of rain had fallen. This was the first night of the cyclonic storms that continued to plague Vanderburgh County and the Ohio Valley for three consecutive nights.

At the blast of the whistle I hurried from the rain soaked makeshift shack towards the clock alley. I splashed through pools of water like large mirrors reflecting the lights of the gantry cranes overhead. Several hundred workers were already in line to clock out.

The three previous nights I had walked the three miles out to the camp, under starlit skies. However, between blinding flashes of lightning accented by distance rolls of thunder, I decided to take the Mesker Park bus. It would take me to within a quarter mile of the N.Y.A. camp. I was allowed to continue sleeping there until I received my first week's paycheck

As the bus neared the top of the hill where it normally stopped at the entrance to the Mesker Zoo, I realized that I was the only passenger left. I pulled the cord to ring the buzzer letting the driver know that I wanted off at the next stop. As he applied the air brakes, I got up from my seat and walked towards the front. The driver looked up into the overhead mirror and gave me a look of disbelief.

As he opened the door, I stepped out into a torrential downpour of rain and booming thunder.

The park was vacant. The howling wind flapped the canvas around the carousel across the street. The blaring bass rhythm of the music maker was silent. The penny arcade was visible only when lit by the brilliant flashes of lightning.

Holding on to my cap, I ran under the stone arch on which was engraved the words, MESKER ZOO. The normally lighted paved pathway appeared dim compared to nature's illumination, followed by the crashing thunder of the storm.

Lightning flashed across the sky as tree limbs swayed ninety degrees to and fro with leaves and twigs filling the air. I felt the rain through my khaki cotton shirt, and my pants and shoes were soaked.

A hundred or so feet down on my right was the tiger arena. Across to my left was a deep pit and the cave home of a roaring lion.

Even though the temperature was above seventy degrees, I felt a cold chill run up my spine, prickling my scalp.

As the storm raged, so did the wild animals with their hideous screams.

In the solid blaze of lightning I viewed a mad king of beasts some thirty feet away in his pit, as he pranced back and forth, roaring above the storm. Suddenly, he made a lunge towards me with his right paw raised to strike. The tiger snarled at me even closer in his enclosure to my right as he clawed at the cement wall that separated us.

The giant wild fowl cage was all aflutter with eerie sounds of excitement caused by the storm. Some of the birds clung to the wire netting while others flew in circles squawking in high tones.

A little farther down the pathway was the Monkey Ship. As I raced pass splashing water, the monkeys were jumping, chattering, and squealing at the top of their lungs. Several were at the top of the ship's mast while others were clinging to the rigging, with a few on top of the captain's quarters.

Breathlessly, I raced on keeping in mind where the snake pit building was located. Frighten by the thought that I was within thirty feet of an eighteen-foot boa constrictor, that could wrap itself around my body and squeeze me to death; my only hope was that he had been fed his five piglets the day before.

My feet could hardly keep up with my body as I ducked swaying wet foliage along the pathway. At times I stumbled until I nearly fell, while trying to keep my eyes scanned for the ten-foot alligator, in case he had escaped from his cage.

At a fork in the walkway, blinded by flashes of lightning, I nearly collapsed from fright as the wind and rain drenched foliage cast-off weird dancing shadows all around me. My jaw quivered as I felt chills running down to my toes.

With the Lord's help, I was able to exit the park along the service road across from the Youth Camp. I was exhausted and wet as a river rat. I had to take a warm shower to remove the shipyard grime and change into clean under clothes before I relaxed enough to go to sleep.

Never again did I enter the Mesker Park Zoo after dark. I know my guardian angel was looking after me that horrible stormy night.

Even now, after these many years, I have goose bumps reminiscing and writing about this experience.

After an extensive search of the July 1942 weather bureau records that are still available, I could find no violent storms recorded during the month. However, if the weather station was located at the Evansville airport, which was approximately five and half miles northeast of the shipyards, it is very possible that stars may had been shinning at that location while storms raged on the southwest side of town.

Chapter

7

Encounter with the FBI

THE FIRST WEEK OF MY EMPLOYMENT AT THE EVANSVILLE NAVY SHIPYARDS TURNED OUT TO BE ONE OF THOSE UNIQUE EXPERIENCES IN LIFE.

Memories stored in the chambers of the brain are connected by a hidden thread of thought only to be aroused as the story starts to unravel like a rhapsody in writing. So it was with the kangaroo court that I endured alone.

For three consecutive evenings during the first week of my employment, severe storms plagued the Evansville area. The torrential rains soon turned the unpaved areas into a quagmire of slushy mud.

Blinding flashes of lightning lit up the sky as ear-piercing cracks of thunder roared overhead. We experienced a rare pleasure inhaling the lightning fresh ozone that filled the air with its clean penetrating odor.

Three of us weld checkers ran splashing through the mud seeking shelter. Lights were on in the upper story of a large warehouse, and the door was unlocked. Clutching our brown bag lunches, we dashed in out of the storm and found ourselves at the foot of a long staircase that was the entrance to the second floor only. Hanging up our raincoats on pegs in the short entranceway, we sat on the stairs and proceeded to eat our lunch.

After the storm subsided and we had eaten, one of the fellows climbed the stairs and hollered, "Hello!" several times. It was apparent that no one was up there.

The second night, as more storms drenched the countryside, we once again visited the unlocked warehouse to get in out of the rain. After eating our sack lunch, we climbed the stairs and ventured to examine the upper story where the lights were on. The entire upper floor was one large room with drafting tables and long wide counter top cabinets with shallow drawers filled with blueprints pertaining to the construction of the boats.

We discovered that the area was called the Mold Loft. We meant no harm, but being young and curious, we examined several of the blueprints.

The third night as the rains continued, we three weld checkers returned to the Mold Loft. After eating our lunch, we once again looked at several additional blueprints.

The following evening as I reported for work on the swing shift where the construction of LST-159 was being built, I noticed that the deck section above the engine room had been welded in place.

With my big mouth I informed Mr. Limeberry, the welding foreman, that the deck section above the engine room wasn't to be welded until after the two 900 HP diesel motors were installed. Evidently, he didn't know about that.

My comment to Mr. Limeberry was like mud hitting a fan. He raised his voice above the ratty-cu-cu-cu-cu of a metal chipper near by and blared out!

"What do you know about it? No one told me anything about such!" Immediately he disappeared.

Approximately four hours later, I was informally introduced to the Shipyard's Management of Authority as a kangaroo court convened on the engine-room deck of LST 159.

A white gloved, Navy Officer, Lieutenant A. F. Sweet, appeared to be in charge. There were also eight other dignitaries that represented authority in some respect pertaining to the construction of the LST boats. A weld checker, by the name of Clendel Williams, was the one to be interrogated!

Although the Lieutenant really wanted the answers to his questions from me, a peon, he went through the ranks, addressing his questions to the highest-ranking shipyard official, Mr. T. R. (Ted) Davis, General Hull Superintendent. In turn, Mr. Davis repeated the question to his assistant, Mr. B. B. Dunville. The questioning proceeded in a trickle down manner through the chain of commands to E.D. Rainey, Superintendent of Welding; Geo. C. Martin, U.S.N.R. Chief Inspector & Naval Intelligence; Roman J. Ritzert, Chief Hull Inspector; plus the three Welding Shifts Superintendents, who relayed the question to Mr. Limeberry my foreman, who in turn asked me the question. The procedure was most frightening as well as confusing, and comical, as I was the only one who could provide the answers that he desired.

I didn't mean to be disrespectful to the Lieutenant in his official position as Judge of the Court. However, I answered his other questions directly without hesitation.

The following afternoon, Mr. Limeberry told me that I had saved the company many thousands of dollars. My reward? I was a spy suspect!

To satisfy my curiosity, the following night, I walked down to the Mold Loft Warehouse Building and checked the door. It was locked, as it should have been all along!

Three weeks later, being somewhat disgusted with the weld-checking job, I did a foolish trick that I could have been fired for. I punched in my time card early, hunted up Mr. Limeberry, my foreman. I slapped him on the shoulder, making certain that he saw me. Then I walked over to the clock alley. I went through the process, acting as if I was clocking out, and walked out with the day shift as they were leaving.

I caught a ride on a city bus to the corner of Franklin and Fulton Avenue. I went to the office of the American Federation of Labor Boilermaker Union, usually referred to as the A.F.L.

I joined the union as a welder's helper. I paid my initiation fee of $20.00 (one week's pay), and they reassigned me to attend the welding school at the shipyards, as a welding trainee, which paid seventy-five cents an hour. They also arranged to have my monthly Union Dues of $2.00 taken out of my paycheck.

The Union boss reminded me of the 1929 gangster pictures. He was very short in statute, with a long gold chain, watch fob, and several rings on his fingers. An oddity with black spats, a black suit, and a large black brim hat. His lady secretary, had rings on every finger, wore a long black satin dress adorned with shiny sequins and long necklaces.

Having received my union membership card and orders to attend the welding school, I loafed around for an hour, ate lunch, and walked back through the clock alley along with several others who had eaten their evening lunch across the street.

As I walked up to Mr. Limeberry, he said, "Where in the hell have you been? I looked all over for you." I showed him the union card and the order to report to welding school the following day.

He said, "Good for you, that's wonderful!"

A fellow named Bob attended the welding school at the same time I was there. At the end of two weeks, we both passed the flat welding test and were certified as tack welders for the ship fitters.

Bob was very talkative. In thinking over our conversations, I realized they mostly pertained to my past education, travels, and family background.

After six weeks, Bob was transferred and I didn't see him again for several months. One evening as I was on my way to the crapper, I saw him with a carpenter hammer hung in the loop on the right side of his overalls. I spoke to him, and he seemed embarrassed that I recognized him. I wondered how he happened to have changed jobs and was now a carpenter helper.

A year later in the fall of 1943 while working on the day shift, I ran into Bob who was operating an acetylene-oxygen cutting machine. I jokingly told him that he must be related to the big boss man to change jobs so often.

Smiling, he said something to the effect that he figured jobs would be scarce when this war was over and was trying to learn several different trades in order to have the experience if needed.

Another year passed, and I was transferred to the graveyard shift, assigned to the skeg welding crew installing drive shaft tubes.

One morning, as I was getting off work and was on my way to the clock alley, Bob met me and said, "Williams, we caught two spies last night on Way 4." Then, as a second thought, he continued, "Please, don't say anything about it. I had to talk to someone. I know that you can be trusted."

So, I thought, that explained how Bob was able to transfer from one job to another. He was a federal agent checking on suspected characters, and I had been one of those that he checked out.

Suddenly Bob's statements raced through my mind like a California wildfire. Had they really caught a spy, or was that another ploy of his to get my reaction when told about someone getting caught? Evidently, I passed my final examination.

In 1971, while on a tour through the FBI headquarters in Washington, D.C. with a group of five other men, we each signed the entrance register.

Along about the middle of the tour, our guide was handled a note from another agent. Reading from the note, he turned facing us and commented. "We have more information on a couple of you gentlemen than you know about yourself."

I replied, "Yes, I'm one of them. One of your agents by the name of Bob checked me out while I was employed at the Evansville Navy Shipyards."

The guide smiled and continued with the tour. It was then that I noticed his I.D. Badge had the name "Bob" on it. Jokingly I asked if he had ever been in Evansville. He replied that he had once passed through there by train.

Secret Agents are good at keeping secrets.

Chapter

8

Money Serial # Game

DURING THE FIRST FEW WEEKS OF MY EMPLOYMENT AT THE SHIPYARDS, THE FOREMAN DELIVERED OUR WEEKLY PAY TO EACH OF US ON FRIDAY.

The exact cash amount after deductions were sealed in a brown payroll envelope with my name printed on it. This was the Missouri Valley Bridge and Iron Company method of payment.

Within a few minutes after counting the cash and feasting my eyes on that ten dollar bill, a five dollar bill, the three one dollar bills, plus the sixty cents in coins, I was asked by a co-worker if I wanted to join their serial number pool for a chance of winning an additional amount of cash.

Being a green kid from the country and not wanting to appear like a dummy, I said, "Sure, how much?"

"Only a dollar." I handled him one of my hard earned dollar bills.

"Let me see your other dollar bills so I can check the serial numbers on them also and see if you might be a winner already. Sorry, but you don't have a winner this time."

"Hey, come back here." Being puzzled about his quick answer I asked, "How do you know?"

"To have a chance to win, you need a full house, a flush or three aces," was his quick reply.

Immediately a winner was announced and all the fun was over.

The following week the winner of the pool last week came by and collected my dollar donation in the hat. That time the winner was my co-worker who had first approached me about joining.

The third week, the winner was the same person who had won the first week.

Still not being too bright, I donated into the serial number pool again for the fourth time. When the winner was announced, it was the man who was the winner the second week!

I thought to myself, now wait a minute! These guys aren't shooting straight. There's some thing fishy going on here and it's beginning to stink, with double winners in a row? No, never again. I can't win when they can do a fast switch with their presorted bills with a full hand of serial numbers while the rest of us don't have a chance!

I heard that crap games had occurred down in the 'B-tanks' and some of the guys went home with hardly enough money to pay for their ride the following week.

About that time, the Missouri Valley Bridge and Iron began to pay us by check instead of cash. That switched the lottery pool over to the check number. There will always be a few that have the betting urge to roll the dice. "Let's have snake eyes," they will yell, "Come on baby! Just one more time!"

Chapter

9

Boarding House Musings

AS SOON AS I RECEIVED MY FIRST PAY FROM THE SHIPYARDS, I HAD TO MOVE OUT OF THE BARRACKS AT THE NYA CAMP AND FIND A PLACE TO SLEEP.

I checked the advertisements of rooms for rent in the *Courier* and finally checked one out on Riverside Drive. The landlady showed me a front bedroom with bath and said, "The rent is $10.00 in advance, please, and is due every Sunday morning."

It was one of those stately white colonial type homes that stood along Riverside Drive before the war. In later years, those homes were torn down and replaced by other buildings.

The bed linens were starched and the furniture was of fine quality antiques. They were a lot nicer than any to which I had been accustomed.

The towels were snow white, and there I was direct from the shipyards in my dirty work clothes. Every item that I touched was soon soiled.

I slept well, but it wasn't like I felt comfortable living there. Everything was too picky spiffy for my existence in those surroundings. Also, I had to eat all of my meals in greasy spoons. I was soon searching for another place to make my abode.

That was when I found the Burger Boarding House out on East Columbia. Ma Burger welcomed me like an old friend and included two meals daily with the price of a bed for $10.00 per week. The food was fabulous, the bed was clean, and the furniture if broken, could be inexpensively replaced. The only major problem was the lack of bath facilities. There was only one bathroom for ten to twelve boarders, providing a lot of door knocking.

The location was two blocks from Briggs & Stratton factory and five blocks from Chrysler and Servel. However, it was three miles from the shipyards, which was probably good as most of the time, it was a quiet neighborhood.

About the only disturbance during the summer months was from an occasional street vendor, jingling a bell while peddling a three wheeled conveyance and shouting, "Ice cream! Ice cream! Six flavors, ice cream!" Then in the winter months, the same vendors would be yelling, "Hot Tamales! Get them while they are hot!"

When I first roomed at Ma Burger's, I was working the swing shift. I would come in about 1:00 AM, take a nice leisurely warm bath and raid the fridge for a piece of pie, after which I ran upstairs for a good night's sleep in an antique brass bed.

Usually I would get up about noon, and Ma Burger would be all smiles and offer to fix my breakfast. I was the only boarder working on the swing shift. In the evening, all of the other men boarders filed in from work to a bountiful feast set on a ten-foot long table in the front room.

Instead of chairs, we sat on long benches while eating our meals. The person with the longest arms usually enjoyed more hot biscuits with real butter and blackberry jelly than anyone else. Most of the time it was a harmonious group, and I don't remember ever hearing a disheartening discussion.

In a group of a dozen personalities there's bound to be one odd ball. Fatso and his Cadillac were there for a short time until he moved out to be with his lady friend, the redheaded nurse.

Then there was Mr. Barr. He worked the day shift at Briggs. "I never touched a drop of alcohol during the week as we are only allowed one rest break during a four hour period. The leadsman has to take your place until you return, so someone else can go to the rest room."

On Friday night, Mr. Barr would come in from work carrying a 24-bottle case of Falls Staff. He wasn't stingy with his drinks as he offered a beer to anyone that would sit down and drink with him.

When I was working on the day shift and getting ready to go out on a date, Mr. Barr would say, "Here Williams, have a beer. You never know how long this night might be." I'd refuse his offer, and tell him to enjoy himself. By the time I returned at near midnight, Mr. Barr would be sitting at the kitchen table holding a half empty bottle of beer in his right hand, with as many as 18 to 20 empties on the table. He would be slaphappy, but he could still shuffle to the bathroom without help. I felt sorry for him, especially his kidneys. The beer bottle seemed to be his weekend companion.

Another fellow that I have tried several times to locate without success was from Paducah, Kentucky. His last name was Gilliam, and he said he was a grandson of the founder of the Gilliam Candy Co. He didn't claim to be the Duke of Paducah, but he couldn't have been any nicer if he had been.

Gilliam went home for Christmas in 1943, and when he returned he brought a large ten-pound sack of hard candy back with him. He shared some of the contents with us fellow boarders, giving the rest of it to Ma Burger for her family and friends.

Sugar was rationed and even the candy companies were limited on the amount of sugar they could purchase. He said, "Some companies making peanut brittle candy used floor sweepings, and if you looked closely, you would find cardboard and even splinters in the candy." Ever since, I check peanut brittle for signs of dirt before making a purchase.

Six fellows from the Salem, Indiana area roomed at Ma Burger's for a short time during 1942. Chester Owens and Bernard 'Boomer' Brown worked at the shipyards before they both joined the navy. Lowell Young, Hager Garriott, and Bernard Baker worked at the Republic Aviation Plant for a while, and I worked at the shipyards.

On a Sunday morning, we asked Ma Burger if there was a Protestant church in the neighborhood. She replied, "There's a Baptist church over on Garvin Street, a distance of about six blocks." Five of us guys walked over there. We filled up one short bench seat. Since we were all of draft age, the deacons felt it was their duty to urge each of us to come forward, confess our sins, and join their church, while they kept singing the invitational song over and over. They had no takers as we were all saved Christians at our home churches.

One time, four Kentuckians asked to room at Ma Burger's. She told them that she only had room for two at that time. They said that was all right. Two of them worked days and the other two worked nights and that they would work out the arrangement. She agreed to give it a try, but she would have to charge each of them the $10.00 room rent.

Everything seemed to be working out quite well for a few weeks. Then, one day when she went in to change the sheets, she smelled a foul odor. There under the bed was a rolled up newspaper full of the stuff. If that wasn't enough, she noticed that the screen in the window was rusted and surmised that instead of them going down to the bathroom, they had been watering the open window screen. That was when it all hit the fan. She packed up their suitcases and set them out on the front porch. When they came in, she read the riot act to all four of them. "I don't rent to hoodlums!"

Ma Burger was nice, but she had her rules and regulations about things. She attended séance meetings where a woman medium claimed to be able to communicate with the spirits of the dead. She talked another boarder and myself, into going with her to one of their Saturday night meetings.

The medium wore a black veil over her head and had a blue crystal ball about the size of a basketball that she worked her hands over and around while mumbling frivolous words that made no sense to me. Finally, she claimed that she had made contact with my grandfather, John Williams who had died long before I was born. What she said was a repeat of what I had told Ma Burger when she had asked me a few days before about my parents and grandparents. The séance was a big hoax.

One morning when I came down for my noon day breakfast, Ma Burger asked if I and three other boarders, who at that time worked the night shift, would go with her to buy shotgun shells. A friend had called, telling her that a hardware store in Owensboro, Kentucky, had received a shipment of twelve-gauge shells. She wanted to get several boxes for her son, Cliff, who was married. Shotgun shells were not rationed but they were scarce.

We all piled into her 1934 Chevrolet 2-door, and off we went with our landlady over to Owensboro, a distance of 43 miles. People were lined up outside the store waiting to buy one box of shells. Ma furnished the money, which was about $1.89 for a box of twelve-gauge shells. After standing in line half an hour, we each purchased one box of shells. Back across the river to the boarding house we rode. Ma thanked each of us for the favor of helping her buy the shells. She had connections and was good at locating scarce items through the black market. Sometimes the boarders would give her a few extra ration coupons for gasoline, which she offered to help others who were in need. We boarders, who ate most of our meals there, gave her our food ration books that included the meat and sugar coupons.

There was that tendency for people to want items that were hard to get, even if they didn't actually need them. It was a form of prestige to brag about to their friends, "Look at what I got!"

Ma Burger's army friend who was a colonel at Fort Knox, Kentucky, had an inside track on the horse races. The colonel had served with her deceased husband during World War I. Every once in a while Ma Burger would say that she had a hot tip on the 'winner' of a certain race that would be run that afternoon, and if any one would want to place a bet through her she would be glad to oblige. I wasn't a betting man, so I never took her up on her offer. If some of the other fellows did, I never heard of it. However, years later, I was told that she had had connections with an off track betting operation at Garvin and East Columbia Street.

The front room of that building was a poolroom during the war. I stopped in there several times while waiting for a city bus and shot a few games of billiards. At the time, I had no idea that it was a front for a 'bookie'. Evidently, some horse races were fixed, with the jockeys being paid off. The winning horse was sure to win unless he fell and broke a leg. It was a neat operation with a direct wire to the track.

To Ma Burger, I owe thanks for more than just a place to lay my head and two meals a day. The knowledge I obtained while living in a boarding house for two years has benefited me so many times down through the years. Such as, "If you are hungry don't find fault with the cook."

Author standing on porch where he boarded for two years while working at the shipyards.

Chapter

10

The Cackling Model 'A'

DURING WWII, MOST EVERYTHING WAS RATIONED. WHAT WASN'T RATIONED BECAME DIFFICULT TO FIND.

In order to conserve fuel and rubber tires, our government slapped on a thirty-five mile per hour speed limit.

The 1930 Model 'A' Ford coupe that Bernard Baker and I bought in August of 1942, cost one hundred and twenty-five dollars. The tires looked to be in good condition with about two-third tread remaining.

Being two silly teenagers, we named the car Katie-Ann, after our girl friends at that time.

I got a small can of yellow enamel and scribed the name Katie on the passenger side front fender and the name Ann on the driver's side fender.

On a Saturday evening, soon after we bought the car, we decided to drive the Model 'A' from Evansville, where we worked, to Campbellsburg, Indiana, where our parents lived, a distance of about one hundred and ten miles.

The eleven-gallon gasoline tank was full. We had given no thought of applying for a gasoline ration coupon book. The fact that the car had neither spare wheel nor tire was of little concern to us.

Without a road map, we headed east out of Evansville towards Boonville. We knew where we were going, so why look at a road map. We were old enough to find our way back to our home base, even though we had never driven the route before.

We stayed on Highway 62, and it was dark by the time we got to St. Meinrad. Somewhere among the hills of the Hoosier National Forest in Crawford County, we had the feeling that we were lost as all the hills and curves in the road started to look alike.

As the Model 'A' cackled around a turn in the road, up ahead was the first real bright spot we had seen since leaving Boonville. We both were thirsty and tired from sitting, so we decided to stop.

A jukebox was blaring away, "You are my sunshine, my only sunshine. You make me happy when skies are gray. You'll never know dear how much I love you. Please, don't take my sunshine away. The other night dear…"

About that time, a pretty little blonde came tripping from a back room over to where we were sitting at the counter and inquired. "What can I get youse guys?"

Before we could answer, she asked, "Where youse guys from anyway?"

Bernard said, "Evansville."

"Oh, my, where's that?"

Bernard laughed, "That's where we are from."

"My, will miracles never cease?" was her giggly reply. "What was it youse guys wanted? We got most anything youse might want."

After she served us our soft drinks, I inquired, "How far is it to Salem?

"Salem? I don't know." She giggled, "Isn't that a long ways from here?"

Bernard jokingly asked to hear what she would say, "Are we in Kentucky?"

She frowned. "I don't think so. Isn't that across the river?"

When we saw the State Road 135 sign at Corydon pointing towards Salem, we knew where we were. We had forty-five miles yet to travel.

We had taken the dogleg route. We should already have been in Campbellsburg as we had been on the road for over four hours and had traveled more than a hundred miles at an average speed of about twenty-five miles an hour.

The courthouse clock was striking twelve, midnight, as the model 'A' cackled around the Salem Public Square.

Sunday afternoon, on the way back, we took a more direct route following State Road 56 to Jasper, and then south on US 231 to State Road 62, taking it into Evansville.

The Model 'A' gas gauge showed empty as we entered the city limits of Boonville.

Not having a gasoline ration coupon, I had the bright idea to stop at the drug store on the Public Square and purchase two quarts of methyl (wood) alcohol. I poured it all in the gas tank.

"I'm glad you thought of that!" Bernard said, with a big grin. "Now we won't run out of gas!"

The old Ford started right off and cackled off the public square towards the west edge of town.

Suddenly the car sputtered, wheezed, coughed, and then after several bang-bangs it died right in the middle of the highway.

There we were. The wood alcohol had mixed with the small amount of gas in the tank, however the mixture was too rich for the carburetor. In trying to start the engine, the battery soon died.

We got out and tried to push the car, and then one of us would jump in and put it in gear. Still it wouldn't fire. Several people stopped offering suggestions; even helped us push to get it started.

Finally, as the sun was sinking low in the western sky, we gave up. We got permission to park the car in the front yard of a family who lived nearby.

Luckily, we were able to flag the Greyhound bus for a ride into Evansville. We told the driver where we lived on East Columbia Street, so he let us off at the nearest location on Virginia Avenue within a few blocks of where we roomed.

Monday morning at breakfast we told our landlady Mrs. Burger, about our misfortune of running out of gas at Boonville.

She said, "Don't worry, I will get you a gas coupon. Then you can catch the Greyhound to Boonville and buy four gallons of gas. That should get your car going again."

Mrs. Burger was as good as her word; however, it did cost me a dollar.

Bernard worked the day shift, so I got a friend that I worked with on the swing shift, to take me to a filling station on the east side of Boonville.

The attendant accepted the black market ration coupon without any questions. He even loaned us his gas can but had another fellow to follow us across town to where our car was parked.

We poured most of the gas into the tank, keeping out enough to prime the carburetor. As long as we fed gas into the carburetor, the motor would start and run.

The fellow that followed us back stayed until the can was empty, then he left. Evidently the station owner had ordered him to bring back the gas can, as it was of more value than the gasoline it had contained which cost 21.9 cents per gallon.

When I was about ready to give up, a couple of guardian angels arrived to help. Two young boys in an old beat up contraption of a street rod stopped and told me to get in the Ford and they would push me down the Boonville Pike until the Ford motor got hot enough to burn the methyl alcohol.

From what was said, they had some knowledge and experience of mixing kerosene and moonshine whiskey with gasoline.

After a mile of being pushed up to a speed of forty miles an hour with many backfires, the Model 'A' took off like it had never ran before. Once it started, I didn't dare stop to talk with them.

Waving my left arm and tooting the horn was my thanks to the boys for their help. That was the last I saw of them.

The Model 'A' rambled off the next fifteen miles in twenty minutes at an average speed of over 45 mph. I did slow down through Chandler. It was a busy wide place in the road during WWII. I was told that they had a town marshal that patrolled part time, and also worked at a war production plant in Evansville.

Guess, I could have been fined for exceeding the national 35 mile per hour speed limit, however, I were so elated in getting the Model 'A' running that I didn't want to give it a chance of dying on me.

I told Mrs. Burger about how a couple of boys in an old jalopy gave me a push for over a mile before the Ford started to run.

"As soon as the motor got hot enough to burn the methyl alcohol, it ran like a scared rabbit, being chased by a pack of yelping, howling beagles."

Bernard was glad that I had gotten the Model 'A' back and parked it in the front yard.

The Cackling 1930 Model A Ford coupe

Chapter

11

A Rough Ride Back

AFTER OUR EXASPERATING EXPERIENCE OF RUNNING OUT OF GASOLINE FOR KATIE-ANN, WE APPLIED FOR AN ALLOTMENT FROM THE VANDERBURGH COUNTY RATIONING BOARD.

According to their regulations our car was not considered an essential vehicle to help the war effort. We were issued a four month 'A' ration book, with 10 coupons, each good for 4 gallons of gasoline.

After supper, on a warm August, Sunday evening, we decided to drive up to Wadesville a distance of about 24 miles. Without a worry or a care in the world, we were enjoying the cool breeze filled with the country fragrance of roasting ear corn and new mown hay.

The Model 'A' was cluck, cluck, clucking along peacefully like an old sitting hen on a batch of eggs ready to hatch. Suddenly, the right rear tire blew out with a nerve-racking bang, as the car veered to the right off the roadway, onto the shoulder, and there we sat with no spare.

Bernard hit the steering wheel with his fist and shouted, "I thought we had good tires. You can drive back, Clendel. I have to work tomorrow!"

"Hum, I also have to work tomorrow."

The right rear tire was flat and there was no road service available. Of course, we couldn't have afforded road service if there had been one.

Bernard jumped out on the driver's side and came around to where I was sitting. "I told you, that you can drive back."

I scooted over under the steering wheel and started the motor, made a U turn across the two-lane highway and started back towards Evansville. The first half-mile wasn't so bad. I drove with the right wheels off the pavement. I was thankful that it was the rear tire and not a front tire that had blown out.

As the evening stars appeared in the western sky, the dew moist air continued to be filled with the aroma of new mown hay, however, with Bernard curled up sleeping in the corner next to the door, it was a whole different world.

We had been talking about old classmates and how we enjoyed our work. Bernard worked at the Republic Aviation plant north of town assembling the P-47 Thunderbolt planes.

Before long, the tire casing started breaking up and lumped crossways on the rim. The car lurched and swayed in jerks, like a ship in rough seas. My head bobbed backward and forward in slow rhythm with the rocking waves.

We covered about three miles that first hour as I clung to the steering wheel. My neck ached as that wheel bumped and flopped along the shoulder. Mile after mile as the tire casing disintegrated, the car finally settled down on the metal rim.

Bernard continued to doze off and on for the next two hours. I suppose, had he been driving, I would have tried to relax and take a nap, also.

As the midnight hour past, a pea soup fog rolled in off the river. The six-volt headlights of the Model 'A' were not of much force even on a clear night. In the fog, I could only see a few feet ahead.

We no doubt awakened hundreds of people as the metal rim flared out to a flat surface, crushing sand and gravel into bits of dust. Suddenly, the noises lessen as the tireless wheel came in contact with a section of smooth cement pavement. The ringing, whirling sounds of the rim were most soothing to my ears.

Near the city limits we crossed several railroad tracks, or that was what I figured they were. The loud metal banging noise, with several hard jolts, added to my neck pains. However, it didn't awaken Bernard. He slept on.

Occasionally, that car rim found potholes in the streets that I had never noticed before. I turned off St. Joe onto West Maryland Street, crossed over on Fulton Avenue to West Columbia Street and then out to East Columbia where we roomed.

It was 1:20 A.M., by the time we returned. Bernard acted disgusted as if it had been my fault that the tire blew out. He went straight to bed, but I had difficulty getting to sleep. Just as I finally dozed off to sleep, he was up ready to go to work.

Since I worked the swing shift at the shipyards, I slept until noon before getting up for a late breakfast. I was sore and stiff, especially my neck.

The next day, Bernard apologized for being so grumpy, saying that he hadn't felt well lately. I forgave him and we remained close friends.

It was a good experience for both of us. We later sold Katie-Ann with three wheels for $75.00. The Model 'A' had its peaks and valleys along with the days when it was parked for lack of gas.

The new owner removed the names Katie-Ann and installed good tires, including a spare wheel.

~~~~~~~TAKE A BREAK ~~~~~~~

A muddled old gentleman met another old fellow on the street corner at Third and Main.

"Hello there!....Why, I haven't seen you in a coon's age. I hardly recognized you. You've changed so much, your hair is different, you seem shorter, and you've done away with your glasses. What in the world happened to you, Charlie?"

"But I'm not Charlie, my name is Jim Jones."

"Amazing----you've even change your name!"

Chapter

12

Casualty on Home Front

I FIND THAT IT IS VERY DIFFICULT TO RELATE TO THE TRAGIC DEATH OF A FRIEND ON DECEMBER 04, 1942.

Double railroads tracks ran the full length of the shipyards along Ohio Street. A freight train was headed west on the outer rails next to Ohio Street, clock alley # 1 crossing. I was one of sixty some swing shift workers anxiously awaiting to clock in before the day shift came bowling out, knocking us down like a bunch of ten pins.

I recognized Charles Owen as the fourth man directly ahead of me. He operated a barbershop in his spare time over on the corner of Garvin and E. Maryland Street. He had cut my hair in recent months. He was a likeable sort of fellow, like an old friend you have known since childhood. His place of business was just a few blocks over from where I boarded. He was also a welder at the shipyards.

As the caboose rolled past, the throng of workers pushed forward across the tracks unaware of an approaching train coming from the opposite direction on the second track.

I was looking down so as to not step on the heels of the man in front when I heard screams.

Everyone came to a sudden halt. Realizing that something serious had happened, I looked up. Someone had stepped into the front of the engine approaching on the second track. The cowcatcher flung a body off to the left clear of the tracks. Blood was splattered on several of the men who had narrowly missed being hit. The dead man was Charles Owen. In an instance, he was gone; nothing could be done to help him.

The thought raced through my mind, "This is a war casualty on the home-front. I might be next." At that moment, the war steeled cold and real, before our very eyes. We were all suddenly shocked.

As soon as the train had passed, the mass of workers, including myself, just as anxious as Charles had been, surged forward past his mangled body lying there. We all needed to get across the tracks and clock in for our war production jobs. Someone would need to take Charles's place on the swing shift as Charles had a new job in Heaven.

An ambulance was arriving by the time I clocked in, so it was only a matter of a very few minutes before help arrived to do whatever could be done for Charles Owen, my friend.

It was a grief that grabs at one's heart. A friend had been taken, however, work must go on.

Jesus said, *"One will be taken and the other left."* (Matt 24:40) King James translation.

A comforting thought is in the following song.

"Work, for the night is coming. Work thro' the sunny noon; Fill brightest hours with labor, rest comes sure and soon.

Give every flying minute, something to keep in store: Work, for the night is coming, when man works no more." By: Annie L. Walker & Lowell Mason.

The construction of the LSTs continued without my friend, Charles Owen. His welding job was assigned to someone else. I missed seeing Charles's pleasant smile, as did his co-workers, his many customers, family and friends.

According to records available, there were nine other home front casualties at the Evansville shipyards during the construction of the LST boats.

There were numerous deaths of war production workers in the Tri-State area due to highway accidents. They were also considered home front casualties of the war.

It was reported that one of the test pilots lost his life while flight-testing one of the 6,700 P-47 fighter planes produced by Evansville Republic Aviation during the war. Also one of the ground crew was killed when he stepped backwards into a rotating propeller.

~~~~~~~~~~~~~~~

"We know not what the morning will bring, but if we will place all of our worries in the hands of Jesus, He will bring us through." Personal opinion of author; Echoes of Freedom.

~~~~~~~~~~~~~~~

As the song states: "*Trust and Obey, for there's no other way but to trust in Jesus and he will bring us through*".

Note: "Trust and Obey" was copyrighted in 1915. Written by Rev. J. H. Sammis and set to music by D. B. Towner.

Passing Thought

WE HAVE NO CONTROL OVER WHAT WE INHERIT FROM THE PAST, HOWEVER, OUR FUTURE WE CREATE.

Chapter

13

Riding With a Drunk on New Year's Day

OUR INTENTIONS WERE TO HAVE A SAFE AND SANE CELEBRATION BY RINGING OUT THE OLD YEAR OF 1942 AND RINGING IN THE NEW YEAR OF 1943.

Bernard Baker and I decided to go to the Carlton Theatre near Second and Main where the Patriotic movie *Yankee Doodle Dandy* was being shown.

Coming out of the theater at one o'clock on New Year's morning we were both energized by the crisp fresh air as well as by James Cagney dancing and singing to the tune of *Yankee Doodle Dandy*.

I was yelling, "Happy New Year!" as Baker yelled, "HEY!" flagging a passing car. "I know that guy. I work with him out at Republic. He'll take us out to our rooming house."

The driver wheeled over to the curb on Main Street. Baker got in the front passenger seat and I climbed into the back seat of the 1933 Plymouth four-door sedan.

I hadn't any more than landed when I inhaled the strong odor of liquor present in the car.

Bernard made a few comments to his co-worker as the old Plymouth sputtered. The driver tromped heavily on the accelerator, flooding the carburetor with gas. After several jerks, the car coughed and then squealed the tires as it took off.

The stoplights for eight blocks on Main Street were centralized. They all changed at the same time. The light at Third and Main was green. The driver shifted through the gears, and the car gained speed.

The Fourth Street stoplight turned yellow as we passed under it. Racing on down Main, the light was also yellow at Fifth Street as we zipped under it.

The Sixth Street stoplight turned red as we approached. The driver gave it no heed and ran the light. On towards Seventh the car raced at fifty miles per hour. That was when I became frighten. This guy was insanely drunk. As I looked through the windshield, I saw the next two stoplights were on red, as bright as Christmas bells on Santa's sleigh.

At Seventh Street, the speedometer needle was pushing sixty, so I slid off the rear seat down onto my knees and hunched behind the front bench seat, braced for an accident.

As we approached Eighth Street at sixty plus, Bernard reached over and turned the switch key off. The driver immediately turned it back on. The muffler blew off with a loud shotgun blast that should have alerted the police. The car roared on without a muffler, as the driver cussed Baker. After two more blocks, he braked heavily and pulled the car over to the curb ordering us both out.

What a joy it was to be able to place our feet on solid ground. Thankfully, Bernard and I would live to see another day of the New Year!

Chapter

14

Bouncing Cars

EVERY SHIFT HAD ITS PROBLEMS IN FINDING ENOUGH PARKING SPACES, ESPECIALLY WHEN A FEW SNOW FLAKES WERE FLOATING AROUND.

There were times when several inches of snow or a coat of ice covered everything. Quite often a vehicle would take up two parking spaces. That would really infuriate the incoming shift workers trying to find a place to park their cars.

In order to make a space wide enough to park another car, three men would grab hold of the rear bumper and three men at the front bumper. They would get the car springing up and down, depending on how good the shock absorbers were. Then on the count of three, they would bounce the vehicle sideways, making room for another vehicle to park.

Most cars manufactured before WWII had bumpers made of heavy strong metal, protecting the vehicle from minor damage. If a car battery was weak, another car could give it a push down the road in order to get it started, and neither car had any damage. Today's car bumpers are fakes made of plastic or rubber. There couldn't be any bouncing of present day cars. An attempt to do so would cause thousands of dollars in damage.

One time, I saw as many as five cars bounced, making an additional parking space. That was done within a matter of a few minutes. There were also times when cars had been shoved so close together that no one could get in their car.

The homebound workers had the biggest problem. Many times they had to bounce cars in both directions before they could get into their car to go home. It was an exasperating ordeal.

The first two years that I worked at the shipyards, I boarded at Ma Burgers, out on East Columbia Street. I usually rode the city bus or rode my bicycle to work. There was a period of time that one of the other welders who drove from the north end of the city offered me a ride to work.

Hank and his wife Sue were both welders at the shipyards. Hank drove a black 1940 Ford opera coupe that had two small seats behind the front seat. On the days I rode with them, there were usually five in the car and we passengers got an ear full of their daily sexual likes and dislikes. Sue's lovely curves were probably twenty years younger that Hank. One morning Sue was fingering Hank's salt pepper curly locks when he told her. "Now quit that, or I'll have to take you back home." "Oh, goodie," Sue giggled. "Then these guys could be our guests."

We were working the swing shift at that time. I participated in the bouncing a few times to help Hank get his car parked. Then at midnight, when getting off from work, we occasionally had to bounce a few cars before Hank or anyone could get into his car.

Chapter

15

Kaintuck Feud Lives On

AN OLD KENTUCKY FEUD DURING THE LAST CENTURY WAS REVIVED AT THE SHIPYARDS WHEN ONE WELDER LEARNED THE NAME OF HIS CO-WORKER.

Jiggs McCoy, a welder on the swing shift over on Way 5, refused to work with his assigned co-worker whose last name happened to be Hatfield.

The foreman asked McCoy if he knew John Hatfield personally. Jiggs answered, "No, I never saw the man before in my life, but I'm not working down in no 'blasted' tank 'wit' no Hatfield."

The feud actually began in 1863 during the Civil War. Most of the McCoys lived in Pike County Kentucky on Tug Fork, a tributary of the Big Sandy River. The Hatfields lived across the river in West Virginia.

The McCoys fought for the Northern Union army while most of the Hatfields joined up with the southern Confederacy. Long after the Civil War was over the fight continued between the families.

The next recorded instance of violence in the feud occurred in 1878 with a dispute over the ownership of a wayward hog. Bill Hatfield found it and Randolph McCoy said it was his pig which had swam across the river. The matter was taken to the local justice of the peace, but since the case was tried in West Virginia, and the justice of the peace presiding over the case was 'Preacher Andy' Hatfield; the McCoy's lost. Two years later, in 1880, Bill Hatfield, was gunned down by two McCoy brothers. They were tried in a Kentucky court and were later acquitted on the grounds of self-defense.

The feud continued with more than a dozen members of the two families murdered. The governors of both states called out the militias to stop the feud and restore order. However, after a sheriff's posse of McCoys made a raid into West Virginia to round up the Hatfields, the Governor of West Virginia accused Kentucky of violating the extradition process by kidnapping the Hatfields. The case was appealed in 1888 to the Supreme Court, which ruled against West Virginia.

The men were tried in Kentucky. All were found guilty. Seven received life imprisonment. The eighth man, "Cottontop" Mount, was executed by hanging. Thousands attended the public hanging in Pikeville, Kentucky. The families finally agreed to stop the fighting in 1891. The last of the feud trials took place in 1901.

Upon further questioning of Jiggs McCoy by the welding foreman, Jiggs replied, "My grandma told me that she was deflowered on a sunny Sunday afternoon under the shade of an old apple tree by a young whipper snapper named Johnny Hatfield, and I remember she had tears in her eyes, and furthermore, he left her pregnant and then married her cousin. I'm telling you, I ain't working with no Hatfield. I'll quit before I'll work with that cuss."

The foreman laughed until he had tears in his eyes, and told Jiggs, "Don't quit. We'll transfer you over to the other end of the yards so you two will probably never see each other again. Then you won't have to worry about what happened to your grandmother years ago. You both are good workers, and we don't want to lose either of you."

Chapter

16

Be Back Soon

THE MOTTO OF SEVERAL OF THE WORKERS WAS: "HAVE GONE TO WORK, WILL BE BACK SOON."

Those words were scribbled on the wall of the outhouse located next to the launching skids and construction Way #1 near the riverbank.

That 10 by 12 building was a busy place with only four stalls and six urinals for thousands of men. It was a place of last resort, and few lingered longer than necessary, as the aroma wasn't conducive to being a loafing place.

A few ornery critters were either too lazy or didn't give a damn to use the facilities. They let it go wherever they were working, especially in the ballast tanks that were called "B-tanks."

Many times the fumes from the arc welding in the bottom of a "B-tank" filled the air with the hot foul odor of urine. I tried not to inhale, however how long can a person hold his or her breath? Occasionally, when excrement was visible, I refused to work until the sanitation crew cleaned it up.

Construction was out in the open. Rain and snow drained into the lower sections of the boat. During cold weather, thick ice formed in the bottom of the "B-tanks."

On cold winter mornings when ice had formed on the steel deck, one had to be more than careful to keep from slipping and injuring oneself. Several times the deck plate would be wet and at the point of freezing. My rubber shoe soles would momentary stick to the metal, making it necessary to take an extra effort to take the next step.

There were times when I was assigned to do pickup welding where it was necessary to lie down on the ice in order to see the place that was to be welded. Usually my heavy clothes, horsehide welding jacket, and leather overalls protected me from the coldness of the ice.

One time my welding jacket froze to the metal and ice. I couldn't rise up from where I had been lying on the ice. As I remember, I had to twist and turn, finally slipping out of the jacket and then use the welding pick hammer to bust up the ice. As I pulled on the jacket, chunks of ice came up with it.

Working conditions varied from day to day and from shift to shift. During the extreme cold weather of winter, it was difficult to know how many pairs of pants and how many shirts to put on. Long johns next to the skin were necessary when the temperature dropped below ten degrees above zero. That was when horsehide jackets and overalls became as stiff as cardboard shipping boxes.

Try pulling on cold leather overalls over three pairs of pants and see how much energy is left to walk stiff legged two blocks to the bus stop or get astride a bicycle and try to ride three miles to work.

I wore whatever clothes the weather called for, and if it turned warm during the shift, I sweated it out. To answer a nature call wearing all those clothes was a problem, especially when the urgent need was pressing.

Then I could relate with the writing on the wall: "Have gone to work, will be back soon."

During the summer months, welders still needed to wear their horsehide jackets and overalls for protection from the welding sparks and hot slag from the welds.

There were days during the summer when the temperature in the ballast or 'B' tanks reached well above 110 degrees with the heat buildup from welding or from the metal shirking crew. This was a sampling of what Hades would be like.

The company furnished salt tablets that were the size of a dime and twice as thick, free to workers. Also, ice water was available.

On those days, perspiration could be wrung out of my horsehide welding jacket and overalls like water from a wet bath towel, and with every step my feet were squishing in my shoes. At that point, I would take a break and down several salt tablets, before I collapsed from heat exhaustion.

Then I might have written something silly like the following, on the outhouse wall.

"I was here but have gone to look for myself. If I should return before I get back, please ask me to wait until some one else takes my place. Have gone to work, will be back soon."

Chapter

17

Johnny Was My Friend

JOHNNY CONANT AND I WORKED ON THE DAY SHIFT FOR ONE YEAR BEFORE HE JOINED THE U. S. NAVY SEA-BEES.

John graduated from the Surgoinsville High School in Tennessee in the spring of 1942, about the same time that I graduated from Salem High School in Indiana.

"Scorchie" was the nickname that Johnny pinned on me, as he couldn't remember my name, Clendel. I had played in the Salem High School orchestra for four years, so upon graduation, they presented me with the school letter 'S' with the abbreviation for orchestra, 'ORCH', across the center of the large letter 'S', which was on a jacket that I wore quite often. Johnny originated the name "Scorchie" from the 'S' and 'ORCH', and even to this day, my wife's family knows me by that name.

Johnny had a sister, age 17, a senior in high school. He was determined that I should meet her. Her picture revealed that she was a very attractive young lady. In the fall of 1943, he asked me to drive him on a weekend to visit his family in Tennessee, a distance of over three hundred and fifty miles. He said he would get the ration coupons and pay for the gasoline, which was 21.9 cents per gallon.

When I looked at a Tennessee road map, I knew that I wasn't going. His folks lived up in the mountains and there were so many uncertainties during wartime. What if we had a tire to blow out or an accident, and not be able to get back by Monday morning, we both would be in big trouble.

One afternoon at work, I told Johnny about a gal I had met on the Greyhound bus the previous Sunday night. The seat next to me was vacant. She inquired, "Is this seat reserved?" I answered, "Yes for you!" For the next fifty miles we had an enjoyable conversation. As the bus neared the Greyhound Bus Terminal in Evansville, she handed me a card with her name, and phone number. I glanced at the card and questioned, "How do you pronounce your first name, 'O-c-e-a-n-n-a?' She replied, "I never liked my given name, 'Oceanna' and my last name is Gosch. Just call me 'O. Gosh', everyone else does!"

Several nights later I called O. Gosch and we met in front of the Lowes Theatre at 6th and Main. She asked me to not come to where she roomed. "It's an apartment unit for girls, operated by the Catholic Church, and Mother Hubbard frowns on us girls having boy friends to call."

That evening we enjoyed the movie, *'Gone With The Wind'*. It was a long movie, about four hours in length. After the show we walked six blocks down Main Street to Hill's for a frosty malt. It was one of those that you could hold upside down and it would not run out. In fact, it was frozen so hard that I bent a metal spoon on my first try.

The following week we met in front of Lowes. That was the best place for us to meet. It was close to the Catholic Apartment where she stayed. O. Gosch said, "I have to be back to the apartment by 10:00 P.M. as Mother Hubbard has placed a curfew on all us girls."

That evening we walked four blocks to the Victor Theatre that was featuring two short movies plus the newsreel, and short run comics.

I kept telling Johnny about O. Gosch. "She is so friendly and talkative that she scares the wits out of me. I need your help in figuring her out. You have been around more than I have. I have a date with her this Thursday night to bowl at the Franklin Lanes. Why don't you meet us there, and I'll casually introduce you to her." Johnny was really interested, "Did you say 7 o'clock? I'll be there."

O. Gosch and I met at our usual meeting place and caught the Franklin Street bus. Johnny already had on his bowling shoes by the time we arrived. O. Gosch took to Johnny just like she had to me. It was just like she had always known Johnny. She was self-poised with her radiant personality, as well as her well-rounded good looks.

The next day at work, Johnny said that in his opinion, Miss O. Gosch, was much older that she appeared. "I would be willing to bet that she is nearer thirty years of age than the twenty she is trying to appear as being."

That was the last date I had with O. Gosch. In the following weeks I tried calling her twice and each time the person that answered sounded like the voice of an older lady, which I assumed to be that of Mother Hubbard. I was told that Miss Gosch was no longer living there.

One weekend I asked Johnny to ride up home with me on a Saturday afternoon and meet my parents. The next morning, Sunday, we attended Bible Study at the Mt. Carmel Christian Church. Johnny appreciated the chance of getting out of the city for the weekend.

Sometime later, Johnny asked me to join the Sea-bees, with him. I told him that I wouldn't be able to pass the physical. The Sea-bees were a Civil Engineer Corps of the United States Navy. They followed the Marines. They built harbor facilities and airfield landing strips for the invasion forces.

Johnny survived the war, married, and settled in Spokane, Washington. Johnny had left in my possession several of his family photos that he had asked me to keep for him until he called for them. We made contact in 1971, and at that time he was in the lumber business with three locations in Spokane.

I never heard directly from Johnny after that. However, in 1991, via a phone call with a man by the same last name living in Spokane, I was told that Johnny Conant had died and that he had no knowledge of Johnny's son or wife.

There are some people you meet during a lifetime that you never forget. Johnny was one of those people for me.

Chapter

18

A Visit To The USO

THE UNITED SERVICE ORGANATION WAS A GOVERNMENT SPONSORED CLUB FOR SERVICE MEN AND WOMEN IN THE ARMY AND NAVY, ALSO KNOWN AS THE USO.

The building was located on 8th Street, just off Main Street, and had originally been a railroad depot. Just before the war it was a Ford dealership.

One Saturday after work, Johnny Conant and I agreed to meet in front of the Lowe's Theater, at Sixth and Main at seven o'clock. We decided to walk over to the USO building and look things over.

The building was packed with soldiers from Fort Campbell and Camp Breckenridge, Ky., both located over a hundred miles away. There were lots of girls our age and older women dancing with the solders as the orchestra played doughboy songs, such as:

"Don't sit under the apple tree with anyone else but me, with anyone else but me, with anyone else but me. No! No! No! Don't sit under the apply tree with anyone but me, Till I come marching home!"

"I told the gang, the whole shebang that you were sweet and true. They ran right out and came right back with a photograph of you."

"So, don't sit under the apple tree with anyone else but me. Till I come marching home!"

Johnny and I stood around for a while, enjoying the music along with free punch and cookies.

Another song that they played and sang went something like this:

" Praise the Lord,
Pass the ammunition, Praise the Lord,
Pass the ammunition, Praise the Lord,
Let's all stay free, so we can soon go home!"

Neither of us had any former training in dance routine, so after enjoying the music and refreshments, we decided we had better leave. However, they started singing the following song, so we stuck around a little longer.

GEE, BUT I WANT TO GO HOME
"The coffee that they give us,
They say is mighty fine,
It's good for cuts and bruises
And it taste like iodine."
Chorus:
"I don't want no more of army life,
Gee, But I want to go home,
Gee, But I want to go home.
Gee, But I want to go home.
"The biscuits that they give us,
They say are mighty fine,
Someone threw one yesterday,
And nearly killed a pal of mine."

“The clothes that they give us,
They say are mighty fine,
However me and my buddy
Can both fit into mine.”

“They treat us all like monkeys
And make us stand in line,
They give you fifty dollars
And then take back forty-nine.”

“The girls at the USO club
They say are mighty fine,
Most are over eighty
And the rest are under nine.”

The building was getting crowded with Army and Navy service men and women, so we left. They were having a ball, dancing to the music by the Shipyard Orchestra.

The sidewalks were packed with service men as we walked back uptown. One soldier stopped right in front of me, clicked his heels and stood at attention saluting. He acted a little tipsy, so I figured he probably had a little too much of the USO punch.

I had on a tan gabardine shirt and matching pants that I had bought from an Army surplus store. The shirt had button down flaps on the shoulders and I had on a black tie.

Another soldier stepped off the walk into the street and turned with a full salute as if he didn’t know whether he should salute or not. He didn’t want to take a chance just in case I was an officer.

I had never given it a thought that my black tie might cause a problem.

Johnny said, "You had better yank that tie off before you get arrested for impersonating an officer."

During the war years, there were lots of wacky songs written. It was the music along with prayers that saw us through those trying times with hope in our hearts. Those songs gave our country a laugh when it needed it most. One wacky song sounded something like this:

Mairzy Doats and Goats eat oats,
And little lambs eat ivy.
I'd eat ivy too, wouldn't you?

Others songs of that day that I'm reminded of were: *Beer Barrel Polka; Bell Bottom Trousers; Three Little Fishes; I'm Looking Over A Four Leaf Clover; Buttons And Bows; If I Knew You Were Comin' I'd Baked A Cake; You Are My Sunshine; Deep In The Heart Of Texas, and The Yellow Rose of Texas*, along with dozens of other wacky songs.

Then there were lots of other dance tunes; *When The Lights Comes On Again All Over The World. The White Cliffs of Dover*, and several Polkas tunes that had everyone jitter bugging.

Then the song we all should be singing today!
"God Bless America, Land That We Love.
Stand beside her and guide her, through the night,
With the Light from above."

Chapter

19

Uncle Sam's Greetings

ON DECEMBER 22, 1941, PRESIDENT ROOSEVELT SIGNED A NEW SELECTIVE SERVICE ACT, "whereby all men from 18 to 64 years of age must register for military service, and all men from 20 to 44 could possibly be drafted for military duty."

I remember writing in my diary on the night of December 31, 1941, these words, "Where will I be one year from tonight?"

The Selective Service Board would draw names out of the pool of registered men to be called up by Uncle Sam; pointing his finger, "I want you!"

I was employed at the Evansville Shipyards, when I received my first draft notice to report for a physical examination the first week of March 1943.

Sixty men including myself were notified that we were to report to the Salem Indiana Armory, at 7:00 AM on Saturday morning, and then travel on two school buses to the National Guard Armory in the capital city of Indianapolis.

Once there, we were ordered to strip off all our clothing including shoes and socks. They had us lined up around the wall in a room where the temperature felt like it was around 56 degrees.

One by one, the doctor checked everyone's blood pressure and pulse. When it was my turn, he told me to stand on my head. I said, "I can't, I never could." He ordered the two naked men behind me, "Hold him up side down for two minutes", while he held a stopwatch.

"Set him down." he barked. He then checked my pulse and blood pressure, looked in my eyes with a blinding flashlight and said, "Get your clothes on."

After everyone was examined, we all climbed back into the buses and were brought back to the Salem Armory not knowing what our future held.

The following week I received a notice that I had been classified 4-F. That meant that the doctor who examined me determined that I was physically unable to serve in any one of the three branches of the service.

The possible reason that I was classified 4-F, was because I was 50 pounds under weight at 118 pounds with a height of six foot, one inch.

I returned to the Evansville Shipyards and continued working as an electric arc welder building LST boats. In one sense it was a relief, and in another, I felt inferior, ashamed of the fact that I wasn't accepted for military service.

Six months later I received another notice from my local 'Draft Board' to report for another military service examination. I took the notice to my welding foreman. The following day, he said they had applied for a deferment for me as I was of more benefit to the war effort as a welder.

Another six months passed and I received notice from my local Selective Service Board that I had been reclassified in 1-A. I was required to take another physical examination for military service.

I told my welding foreman who advised me to respond to the draft board requirements, but he felt that I would be back on the job within a few days.

As in a bad dream, the same scenario was repeated. On a frosty Saturday morning in March, sixty men met at the Salem Armory and were bused in two school buses to the Indianapolis National Guards Armory.

Once more I found myself standing around the walls of the armory with fifty-nine others, naked as a 'Jay Bird', in shivering temperature on a windy March day. It wasn't very pleasant and certainly no fun. After half an hour of shallow breathing and trying to hold my breath to keep warm, the same old doctor walked past me, stopped, turned around, and declared; "You again? Get your clothes on. You're not going anywhere."

Although I had gained ten pounds since the previous examination the year before, I was still a beanpole at six-foot-one. The actual reason for my being so much under weight was because of an open fractured skull I received on March 18, 1938 in high school physical education, according to my doctor, Dr. Lawrence Paynter, of Salem. He said the head injury had stunt my ability to gain weight.

It was incredible that the Army doctor should remember me from the thousands of inductees he, no doubt, had examined during the previous year.

My walking skeleton evidently, had made quite an impression on his photographic mind.

So I gratefully returned to my welding job at the shipyards where I continued working for the next sixteen months until the last LST boat was launched.

Chapter

20

Crowded City Buses

THE EVANSVILLE TRANSIT SYSTEM WAS OVERBURDENED WITH THOUSANDS OF SHIPYARD WORKERS WHO DEPENDED ON CITY BUS TRANSPORTATION EVERY DAY.

Most of the buses were designed to seat 28 passengers, but usually twice that number would crowd in to be transported up and down Franklin Street on the west side of the city near the shipyards.

Sometimes there were so many people crowded in the front entrance that the driver would have to ask several to step off before he could close the door. Some riders exercised their freedom to say and do as they pleased without concern for their fellow riders. At times, a layer of blue cigarette smoke would be so thick overhead that the advertisements above the windows couldn't be read.

One evening, a half dozen city buses were lined on Franklin Street headed for downtown. I managed to crowd onto the first bus. A shipyard welder, smoking a cigar and wearing a welding hood on backwards, was taking up two seats. I bumped him over as I sat down. There was such filthy language, calling me names and telling me where to go! It wasn't a 'he'; it was a 'she'.

I got up, pulled the 'rip cord', and got off at the next street corner. I was nearly suffocated as the bus took off, leaving me behind in a cloud of smoke. I later caught another bus, paying my fare again.

Those were the days of a free education. I grew up fast, learning words and names of things that I never heard of before, nor knew existed.

One nice day when I had nothing better to do, I boarded a city bus with the street name of Heidelbach above the windshield. I paid my fare and rode for the next hour. At some point, the bus made a loop around a city block and headed back toward downtown. I saw sections of Evansville I hadn't seen before or since. When the buses weren't crowded, they provided a cheap form of entertainment. One just needed to sit back, relax, and enjoy the ride.

By riding the bus every day, faces became familiar even though names were not exchanged. Casual conversations were often shared, but most of the riders I never cared to acknowledge.

One fellow struck up a conversation, but I wasn't paying much attention until I heard him say something about being an auto body repairman. Since I was interested in that line of work, I listened closely as he rattled on about a young lady wanting her car painted. He continued, "I went to make an estimate on taking out a few dents and painting the car. She invited me in to look at a magazine with a picture of a car that showed the color she wanted her car painted. While I was glancing through the magazine, she excused herself for a minute and said,

'I'll be right back.' And she was. She landed in my lap, with nothing on but a sheer pink negligee, and started smothering me in kisses. I think she was trying to get a free paint job."

At the next block was where I got off, so I don't know how he ever came out about painting her car. Thank goodness, that was his dilemma, not mine!

Riding a bus during the war years was the best method of transportation in the city, and it was very economical. As I remember, Evansville Transit fare was a dime per ride. They offered three tokens for a quarter. For one token you could ride to anywhere that bus traveled within the city plus a transfer ticket to continue your ride on another bus if needed. Plus, the entertainment was free!

~~~~~~~~TAKE A BREAK ~~~~~~~~

Two elderly gentlemen were sitting next to each other in the doctor's waiting room.

"I'm aching from arthritis," said one.

"I'm Ben Bent from Birmingham," said the other. "Certainly good to know you."
~~~~~~~~

Chapter

21

Old Blue The Foreman

THE TEMPERATURE WAS COURTING THE ZERO MARK AT A LITTLE PAST EIGHT ON A COLD, BLUSTERY FEBRUARY MORNING.

Old Blue's welding crew was made up of ten men and two women. They were gathered around a hot, cherry red, improvised stove to keep from freezing. Graveyard welders had fabricated the makeshift stove during the early morning hours.

Most likely the major panels in the improvised stove were intended for use in the construction of the LST boat that we were building.

Several wooden skids had been busted up for kindling to start the fire. A scavenger hunt for more wood had provided sufficient fuel to keep it burning, providing some warmth from the bitter cold.

According to the *Evansville Courier* and the *Associated Press*, the extreme cold weather that had suddenly dropped out of Canada was a Siberian Express.

The icy blast was flapping the tarps covering the two by four framework that had been erected on three sides, four feet from the stove. Everyone was enjoying the warmth until Old Blue staggered up.

No, he wasn't drunk but was staggering under the weight of two five-gallon buckets, one in each hand, filled with coal. He had probably swiped the coal from down on the dock where they operated a steam drilling rig. He emptied both buckets into the top of the stove, nearly putting the fire out.

Everyone had been fairly comfortable until Old Blue dumped all of that coal into the stove. Then he pulled up a long high-pressure air hose, sticking it into the bottom of the stove.

The stove immediately became a gigantic hissing blowtorch, belching flames with dense smoke shooting from the six-foot chimney. The cherry red sides soon changed to a lemon yellow as heavy black smoke swirled down off the deck to ground level.

The dark red flames smothered down for a moment. Then with an explosive force, fire shot high above the windbreak and the entire stove vibrated as if it was going to take flight.

Everyone scampered out into the cold as the stove wheezed, puffed, and then roared! The tarp four feet from the intense heat started to smoke.

The 3/8th., inch boilerplate sides warped. The smoke stack keeled sideways as it all melted down.

About that time, one of the safety-men with the green cross on the front of his hard hat, ran up the steps and yanked the air hose out of the bottom of the stove, yelling in a loud demanding voice, "WHO'S THE IDIOT THAT STUCK THE AIR HOSE IN HERE? YOU'RE GOING TO BURN THE YARDS DOWN!"

Soon the fire department arrived and put the fire out. A gantry crane moved into position to remove the molten chunk of iron that had recently been our source of heat. Without the windbreak, the icy wind seemed to intensify as it whirled off the riverbank.

We all suffered from Old Blue's antics. No one dared to say a word, and he certainly wasn't going to admit that he was the one that stuck the air hose in the stove. Cold and shivering, we stood around murmuring about "Old Blue."

If you are wondering why we called him "Old Blue", it is evident that you have never been around anyone that cussed until it was like a contagious disease affecting everyone, or you wouldn't ask.

He was from the oil fields of Oklahoma. Each time he opened his mouth a string of cuss words came out. His speech was so heated with profanity that it literally turned the air blue.

Another one of his personality traits was a violent rage that he didn't seem to be able to control.

He laughed about the time he had a new 1937 Chevrolet Drilling Rig Truck stuck up to the axles in the red gumbo of Oklahoma. Old Blue bragged about hitching a D-6 Cat to the front of the truck, dragging the front half out, leaving the bed with the drilling rig and rear end still stuck in the mud.

He was in charge of roll call each morning. Old Blue had our names listed alphabetically in his little black book.

He didn't have a decent word for any of his workers, cussing us out while assigning job locations. All of the men were s.o.b's and no one ever heard him call a woman worker, "sweetie" or "honey." Slut and Bitch were the norm greetings.

I soon realized that if I was going to work for this cussing devil, I had to let what he said go in one ear and out the other. The environment of just being around him was getting to me. In order to avoid becoming like Old Blue, I had to ignore what he said as much as possible.

Evidently, he noticed that I wasn't paying much attention to what he had to say. One morning, after having assigned work detail to everyone else, he turned to me and yelled, "Willie, what in the "blankety-blank" do I have to call you to get you to cuss me back?"

I replied, "What good would it do?"

That really infuriated him. He jerked off his black welding cap, threw it down onto the deck, and jumped up and down on it in a childish tantrum.

He yelled, "Get the blankety-blank out of my sight!"

I asked; "Where am I supposed to work today?"

Angrily he screamed; "Go back to where you were yesterday!"

"There's nothing left to do down there." I answered. "I completed all of the welds that were marked yesterday."

"Get the h... out of my sight and stay hid!" he shouted.

Given Old Blue's blessings I spent part of the morning napping down in the ballast tank where I worked the day before. The rest of the day I shot the breeze with several of the other welders.

Several weeks later I realized that Old Blue just wasn't his usual self. Very seldom did he utter a cuss word. Apparently, he was not feeling well, and he had very little to say.

Each morning during our roll call and job assignment, he didn't cuss anyone out.

Everyone noticed the calmness and softening of our boss man from the brutal oil fields of Oklahoma. However, no one dared to mention or ask why he wasn't cussing like before.

A week or so later, after he had given everyone else their work assignment, he asked me in a mild-mannered tone, "Have you noticed any difference in me lately?"

Smiling with a half chuckle to myself, I said, "Yes. You aren't cussing like you used to."

He went on to ask, "Do you remember sometime back when I asked you why you didn't cuss me back like the other workers?"

"Yes," I replied.

Old Blue continued, "Your answer was, 'What good would it do?' That got me thinking, and for some reason your reply hit me right here." He pounded on his chest. "Right then I decided to try to stop my cussing, and now my wife thinks that I'm sick!"

The "new" Old Blue continued. "The other night she asked if I felt bad, and I told her that I had never felt better in my life. She doesn't believe me! She said the past few weeks have been the most pleasant time in our fifteen years of marriage!"

~~~~~~~ TAKE A BREAK~~~~~~~

Three rather deaf old geezers met as usual for their morning coffee break and joshing.

"Windy, isn't it?" Said one.

"No, we're at Abry's, not Wendy's, corrected the second old-timer.

"No it's Thursday," replied the third oldster.

"So am I," said the first man. "Think I'll have a glass of lemonade."

The second fellow who worked part time at the hospital said. "I see they've brought in another case of tonsillitis."

Third oldster replied, "Anything is better than that lousy lemonade they've been serving lately.

"Since you mentioned it," said the first man, "Think I'll have a slice of lemon pie with my coffee."

Chapter

22

Fearless Old Blue Learned A Lesson

A FEW WEEKS LATER, THE WELDING MACHINE THAT I WAS USING HAD TO BE MOVED BECAUSE THE WELDING CABLE WASN'T LONG ENOUGH TO REACH MY ASSIGNED WORK AREA.

Old Blue was going to help move the machine and hook it into a 440-volt electrical panel that was seven feet tall, three feet wide, and a foot thick. The high voltage breaker box had receptacles for 12 welding machines. I called it a Pill Box.

Blue yelled, "Come on Williams. Give me a hand here, while I plug this cable into this pill box."

I yelled back, "Blue, get an electrician to plug the machine in. That's what they are supposed to do. Now don't go messing with 440-volts! You'll get yourself killed...you'll never know what hit you!"

He laughed and yelled, "Williams, . . . just a few more feet. I'll plug it in. We don't need any electrician. I can do it! Just a little bit more!"

As I pulled on the electric cable, he socked the cable plug into the 440-volt electric control panel outlet. It sounded like a sonic BOOM!

KkkkkkaaaaaaaaZzzzzooommmBooooooomm!

A blinding, shimmering halo of blue/white high voltage enveloped the entire electrical panel. A white cloud of fumes rose high into the air. Electricity on the entire Way was knocked out.

It's a wonder that Old Blue wasn't electrocuted. It just wasn't his time to go.

I had stumbled backwards over a stack of iron pipes three feet high, landing on my shoulders with my welding boots sticking straight up in the air.

From where Old Blue had fallen on his big fat rear, all that he could see was my feet sticking up.

I heard him climbing over the pipes, trying to get to where I was lying. I wasn't hurt, but in the position I was in, it was impossible for me to get up as my head and shoulders were flat on the deck.

He couldn't see my face as the welding hood had fallen down over my head. I heard Old Blue utter a frantic cry.

"Oh, Lord, what have I done? Oh! No! Willie! speak to me! Say something! Please!"

I was laughing under my hood. Here was the rough and tough former oil rigger so concerned about my well being and he hadn't uttered one cuss word. If such had happened a few months earlier, his cussing would have produced more smoke than the 440-volt explosion!

As I lifted my welding hood, I was startled to see Old Blue's hair and face covered with a gray-blue ash powder, a glimmering ghost of Halloween Past, as he gasped for air. He expected to find me either dead or severely injured. Our guardian angels had saved us both from serious injury.

Immediately, the top deck was swarming with half a dozen electricians. Also, four white suited Safety Men with green cross emblems on their white hard hats came carrying a stretcher and an emergency oxygen pack. They were prepared for the worst possible disaster in case someone had been injured.

Before long, several dozen workers, including ship-fitters, metal-shrinkers, tank-testers, welders, out-fitters, and sanitation workers, came crawling out of every section of the darkened hulk of the ship.

There was no electricity on the entire Way and it was over an hour before power was restored. The electrical panel had to be replaced by the gantry crew while the electricians worked like beavers replacing a transformer and checking the fuses in the replacement panel, making certain that everything was back in operating condition.

Old Blue had been at it once more, with his superior know-it-all attitude, while actually lacking good common judgment.

I forgave him, as down deep he wasn't all that bad a person. His attitude had improved in many ways as he had gotten control of his temper and foul language, at least around me.

Warning To Shipmates

If everything seems to be running smoothly; Obliviously, something has been over looked.

Captain Scorchie

Chapter

23

Peddling Around Town

MANY TIMES THE BUSES WERE SO CROWDED THEY WOULDN'T STOP TO PICK UP ANY MORE PASSENGERS.

So I decided to solve that problem and buy a bicycle that was on display at a Firestone Store.

The clerk said, "I'm sorry. I need to make a sale, but I can't sell you that bicycle unless the ration board approves your application."

That seemed odd to me. I had the money, and he refused to sell me the bicycle. Wasn't this a free country?

Not really, with a war going on. Bicycles were rationed the same as automobiles, gasoline, and shoes. Nearly everything was either rationed or unavailable. President Roosevelt, anticipating that the war would cause inflation, set up the Office of Price Administration to hold prices and wages at their current level.

The nation was at war and everyone needed to abide by the rules and regulations to help the war effort. It was a war that we intended to win and did win through determination, prayers and sacrifice.

The clerk handed me an application form and I filled it out. He sent it to the Vanderburgh County Ration Board for their consideration.

To my surprise, my application was approved. There was no sales tax in Indiana at that time and for the price of $34.95, I rode off into the sunset one warm February day in 1943. I registered the bike at city hall, paid for a city license, in case it got stolen.

The fact that I was employed in defense work and lived three miles from the shipyards was most likely the ration boards deciding factor in granting my request. The bicycle provided needed transportation as well as a lot of enjoyment. My normal route to the shipyards from East Columbia Street, where I boarded, was to ride due west on Columbia to Fulton Avenue. I jogged over, half a block north, to West Maryland and then west to St. Joe Street. I turned south on St. Joe to Ohio Street where the shipyard clock alley # 1 and # 2 were located.

One warm summer night as I got off work at midnight, I sped toward Main Street at Dress Plaza, I heard whistles and noticed the silhouette of a thinly clad lady of the night standing in a lighted doorway. I put all of my energy on the pedals and no doubt the bicycle tires were raising dust. I never rode up Ohio Street again at night, as I didn't consider it a safe area to be in.

On my way to work one afternoon, as I rode down West Maryland Street, an approaching car veered toward me. I steered to the right off the pavement. The car front bumper hit the rear hub of the bike, knocking me about six feet sideways.

I stayed astraddle of the bicycle, but my hard hat rolled off some twenty feet away. I yelled, "What do you think you're doing.... you idiot!"

From the glassy look in his eyes, he had to be under the influence of something. He just stared at me and didn't say one word. A woman and several children in the car had that same startled look.

My guardian angel was riding along with me, because I wasn't hurt, only riled up. I was disgusted to think that anyone would deliberately try to hit a person riding a bicycle. I had the right-of-way and was in plain sight. The driver was in no condition to be driving. Maybe he had fallen asleep, although my yelling should have awakened up the dead!

If it were raining or snowing I would try to catch the city bus, however, I rode the bicycle to work most of the time.

It rained one March night before I got off working the swing shift. Riding up St. Joseph to Maryland I had no problem. However, as I crossed over on Fulton Avenue to West Columbia Street I ran into ice. The bicycle rear wheel wanted to go right as the front wheel attempted to go left.

Needless to say, that was an experience that I well remember. Forward progress was almost impossible.

The street was a glare of black ice, so slick that I was safer on the bike than trying to walk and push the bicycle. I managed to get over to the edge of the street next to the sidewalk where there were a few bits of sand and pea gravel that provided some traction. As I crossed over to East Columbia, the ice dissipated. The last half dozen blocks were ice-free. One or two degrees in temperature made the difference between ice and no ice.

A year later I was working the day shift at the shipyards. One evening after a good supper of steak and gravy, beans, taters, and a big slice of Dutch apple pie, which Ma Burger was noted for, I was feeling energetic.

I straddled my bicycle, rode down Lincoln Avenue, which was also US 41, and into the Commonwealth of Kentucky without having to cross the Ohio River. You might wonder how that is possible, as it was a surprise to me also.

After some checking, I was told that back in 1792, the northern boundary of the state of Kentucky was declared to be the low water mark on the north shore of the Ohio River. Then, in later years, the river changed its course, slicing off a chunk of land, leaving it north of the Ohio River, which is called Green River Island. That is where Ellis Park Race Track is located. Carnivals are allowed to set up there and neither the City of Evansville nor the State of Indiana has any control over them, as they are legally located in Kentucky.

The Evansville city limits extends to the south side to Eagle Creek, which is a marsh land next to Green River Island. Then it is about another mile to the Audubon Memorial Bridge over the Ohio River.

The temperature was in the low seventies with a musky river odor lingering on the evening breeze. Otherwise, it would have been a perfect evening for bicycle riding. I rode past the entrance to the Audubon State Park and was soon in the city of Henderson. According to my watch, it had taken an hour and half to ride the estimated 15 miles.

The sun was setting low in the western sky, as I turned the bike around and headed back towards Evansville. I realized that I had probably taken on more of a ride than I had anticipated. My legs were feeling the effects of pain with each rotation of the bicycle pedals.

Darkness soon encased the landscape as puffs of fog, like white smoke, rose steadily from the murky waters below the desolate bridge. The return trip seemed to be three times longer as dampness of the evening crept in. There was no moon, but millions of stars in the Milky Way were visible. I had the feeling of loneliness as the low drawn out moans of a foghorn sounded from down river.

As I entered the city, the streetlights blinked through the tree leaves as they swayed back and forth on a gentle breeze. By ten o'clock I was back at the boarding house and ready for a shower. Before retiring, I had to have another slice of that delicious Dutch apple pie of Ma Burgers.

Although the bicycle had no odometer, I figured that I had ridden close to 30 miles. Possibly due to gasoline rationing traffic was sparse, as during the entire trip, I encountered no more than a dozen vehicles.

On another afternoon, I remember riding out on US 41 to the Republic Aviation Plant where they assembled the P-47 Thunderbolt fighter planes.

A test pilot was going through the process of putting a new plane through its routine testing. I watched and listened as the plane made a vertical climb, disappeared, and stalled out. Several seconds passed without a sound from the plane. Then, a whistling sound grew louder and louder as the plane plunged earthward. At what seemed to be the very last second, the pilot pulled the plane out of the dive and once again shot vertically skyward with screaming pulsation.

I understood that the test pilot received $50.00 for each plane he tested. That was considered big money during the 1940s; however, one mistake or one defective plane and it would be certain death. It was reported that one test pilot was killed while testing the P-47 Thunderbolt fighter planes.

After Edna, and I started going steady, I often rode my bicycle the ten blocks from my boarding house on East Columbia, to where she roomed on Grand Avenue, near Lincoln Avenue.

We usually caught a bus to go downtown to the theater. Other times we walked as there was no other means of transportation other than the yellow cabs, and I was too stingy to pay the cab fare. What a stinker I must have been back in my earlier days, and I probably haven't changed much since. The experience of growing up during the depression years of the 1930s left an indelible urge to be conservative. I'll likely carry that trait to my grave. No need of flowers at my funeral. Save your money and buy my books is all I'll ever ask of you.

Chapter

24

Rev. Reveals' Mission

IN AUGUST 1942, LOWELL YOUNG AND I ROOMED AT MA BURGER'S BOARDING HOUSE ALONG WITH TEN OTHER MEN.

One Saturday afternoon we started to walk downtown and decided to take a short cut. We didn't know the city, but from our observation, one block of Locust Street appeared to be a red-light district, although salvation was available next door.

At 216 Locust Street, we couldn't help but notice a new looking 1941 Ford station wagon with the hard maple paneled body parked in front of one of the buildings. On the door panel was a decal, Rev. Reveal Mission. Then we noticed the building had a sign, REV. REVEAL MISSION above the door.

The organ music attracted our attention. Evidently, the Lord guided us, anyway we went in. The reverend was delivering his sermon to five men.

"FOR GOD SO LOVED THE WORLD THAT HE GAVE HIS ONLY BEGOTTEN SON, THAT WHOSOEVER BELIEVETH IN HIM SHOULD NOT PERISH, BUT HAVE EVER LASTING LIFE." John 3:16 *King James version*

"Come on in fellows. The Lord welcomes you to the Reverend Ernest I. Reveal Mission. Come on up front and have a seat."

Evidently, our entrance gave the speaker renewed energy, as he pulled out both barrels and shot from the hip.

Reading from Romans 1:16, 17. (King James)

"For I am not ashamed of the gospel of Christ: for it is the power of God unto salvation to every one that believeth; to the Jew first and also to the Greek.

For therein is the righteousness of God revealed from faith to faith: as it is written, the just shall live by faith."

"Fellows, you are looking at the lowest down, worst sinner that ever walked on the face of God's green earth. You name it and I have probably done it. A few years ago I was riding the rails. I was a bum, a beggar, not knowing where my next meal was coming from. There were times that I didn't eat for days. I stole anything I could get my hands on in order to eat. I was the king of the hobos if you want to know. We would build a fire under a railroad viaduct down near a creek and heat what food or coffee we had in a tin can that we found where someone had tossed it. I slept, covered with newspapers trying to keep from freezing."

"THE DEVIL WAS IN ME....HE HAD ME IN HIS GRIPS!....I WAS HIS! ....HIS FIERY FURNACE WAS WAITING FOR ME. THAT WAS TO BE MY REWARD FOR FOLLOWING THE DEVIL! HE WAS READY TO THROW ME INTO THAT ETERNAL LAKE OF FIRE, THE

UNQUENCHABLE FIRE OF HELL, THE FIRE THAT NEVER GOES OUT!'

"IS THAT WHAT YOU WANT, BROTHER? DO YOU WANT TO BE TORMENTED AND BURN FOR EVER, AND EVER, AND EVER?"

"BROTHER, JUST BECAUSE SHADRACH, MESHACH AND ABEDNEGO SURVIDED THE FIERY FURNACE WHEN OLD KING NEBUCHADNEZZAR THREW THEM IN IS NO SIGN THAT YOU ARE GOING TO SURVIVE...."

"THEY HAD THE LORD IN THERE WITH THEM..........YOU NEED THE LORD NOW!"

"I FOUND THE LORD! I FOUND JESUS! HE AND HE ALONE SAVED ME FROM THE BURN PILE!" (Invitation music softly began.)

"DO YOU WANT TO GET SAVED? TODAY IS THE DAY.......TOMORROW MAY BE TOO LATE!....THE LORD IS WAITING FOR YOU TO ACCEPT HIM NOW. HE will forgive you of your sins if you will only confess and ask for forgiveness."

"COME BROTHERS WHILE WE SING: ***'Just now your doubting give o're, Just now reject him no more, Just now throw open the door and let Jesus come into your heart'..."***

"BROTHERS WHY WAIT? THIS MAY BE YOUR LAST CHANCE. THERE'S NO PROMISE OF A TOMORROW............You might be called tonight...WHO KNOWS? Only the Lord knows when your number will come up."

"TODAY IS THE DAY THAT THE LORD HAS MADE. MAKE THE GOOD CONFESSION NOW, AND BE FREE OF YOUR SINS! Why wait any longer, Brother? ACCEPT THE LORD'S INVITATION NOW! TODAY!........Let's pray."

The other men evidently were homeless, staying there at the mission, as Lowell and I were the only ones leaving.

The Reverend was at the door and asked us half a dozen questions while he handed us a small pocket size autographed New Testament.

When I asked Lowell, after sixty some years, if he remembered getting the New Testament, he replied. "Not really. If I got one, I gave it away some years ago along with several other New Testaments that I gave to the Boy Scout Council."

I still have mine. That was the only way that I could have remembered that afternoon so vividly some sixty-nine years ago.

Some will reject the above type of preaching. There was no mention of baptism or immersion to wash away one's sins. Others will say that Jesus accepted the thief on the cross on his confession and through the grace of God and His son, Jesus Christ, all things are possible...So the above sermon has its good points and possibly the Rev. Revel was offering a first step message and had a second step message to deliver to those that accepted Christ.

That brings to mind the story of the man that went up to the preacher after the service and said, "You didn't have to cram the whole load of hay down me on my first visit!"

Chapter

25

The Man From Ketchum

THE EVANSVILE SHIPYARDS PROVIDED EQUAL OPPORTUNITY EMPLOYMENT TO EVERYONE WHO WANTED TO WORK.

That was one of the reasons why a man from Ketchum, Idaho, was working at the shipyards as a courier when I first met him.

This man, named Pedro Crawford, had an aunt living in the area. She had written to him encouraging him to find employment in Evansville.

Pedro was an amputee with only one arm. I hated to ask him how he lost an arm, but I noticed that he rolled his cigarettes as quickly as anyone could with two hands.

Back in the days of the early thirties, a lot of smokers had the nicotine habit but couldn't afford a quarter for a pack of cigarettes. They would buy a sack of Bull Durham smoking tobacco, which included a wrapper of 24 cigarette tissues for a dime, and made their own smokes.

It was fascinating to watch Pedro roll one. He did it so efficiently that it is difficult to describe the process he used. He kept his tobacco sack in his left back pocket, cigarette papers in his left shirt pocket, and a small box of matches in his right shirt pocket.

One day, I asked Pedro if he could show me how he rolled his cigarettes.

"Well," he said, "I'm not much of a teacher, but I'll show you how I do it."

So I squatted down in front of where he was sitting and watched as he reached in his left rear pocket, pulled out his tobacco sack, took one end of the draw string between his teeth, and with his left index finger, made an opening in the top of the sack.

Still holding the sack with his teeth, he reached in his left shirt pocket and pulled out one cigarette tissue paper, which he clutched between his index and middle finger, using his thumb to make a groove in the tissue between his fingers in which to place the tobacco.

Pedro placed the grooved tissue in a fold of his pant leg, took the sack, and tapped out just enough tobacco to fill the groove. He then tightened the drawstring with his teeth and stuck the sack back into his hip pocket. Next, using his index and middle fingers, he gradually rolled the tobacco filled tissue with the aid of his thumb.

I may have gotten so engrossed in his conversation that I missed part of the procedure, but that is the way it looked to me.

I asked, "How long have you been rolling your own?"

"Five years."

"Was that when you lost your arm?"

"Yap," he answered. "I was on a ski patrol rescue run and got one ski caught in an ice rut throwing me into a small fir tree. My right arm was so mangled they had to amputate. It's a miracle that I survived."

Then with his thumb, he smoothed the tobacco out even in the groove, using his index and ring finger to roll the tissue up, and licked the edge length-wise with his tongue to seal it. Sticking the cigarette in his mouth, he struck a wooden match on the seam of his blue jeans and lit up, taking a draw.

"Were you employed at Sun Valley when you had the accident?" I inquired.

"Yes, at the time," he replied. "They kept me on as an emergency radio operator until I quit and came down here to Evansville."

"You said that you were from Ketchum. Just how far is Ketchum from Sun Valley?" I asked.

"Yes, It's right there, hardly a mile apart. At one time there were several lead and silver mines in the area. I did a little prospecting myself but never found anything. The Union Pacific Railroad helped to promote the area as a health retreat as the hot springs furnished relaxing baths for those that could afford it. Then the moneyed folks came in and made it into a big ski resort, naming it Sun Valley"

He continued, "I like it down here. Indiana weather is a lot milder than the rough winters we had in Idaho with blizzards and 20 to 30 degrees below zero temperature. I like to fish and I understand there's lots of good fishing spots in the Tri-State area of Kentucky, Illinois, and Indiana."

Working with fellows like Pedro made life interesting as well as educational.

This certifies that

Clendel F. Williams

of the

Evansville Shipyard

The Missouri Valley Bridge & Iron Co.

(Shipbuilding Division)

Evansville, Indiana

has by loyal effort and efficiency contributed greatly to the winning of the Army-Navy Production Award for Excellence in War Production

NOVEMBER 28, 1943

GENERAL MANAGER

The Army -- Navy "E" Award

The hope and expectations of some of the shipyard workers, and a total surprise to others, was the announcement that the Army–Navy "E" Award had been granted to the Evansville Shipyard.

On November 28, 1943, more than 15,000 persons attended the official award ceremonies, which were held at the shipyards, from the platform at the launching way. At that time, The Army – Navy "E" Flag was presented

This significant award, not easily won nor lightly bestowed, made on a continuing basis to the men and women of the Evansville Shipyard, who by their attitude, example, and devotion to their duties, proved themselves eligible to receive such honor.

All employees received a certificate and also a small "E" Flag pin.

John Conant, Howard Hubele, and J. P. Hulvershorn, three of my co-welders, each received an additional citation, as they individually made a suggestion to the Job Method Improvement Department in which to improve welding methods in certain construction areas of the LSTs.

The Evansville Shipyard constructed the following ship number groups. However, the build schedule was not sequential to the ship numbers as LST-157 through LST-180 were the first ships built. The first LST-157, took over 6 months to complete. By the end of the war, it only took 65 days to build a LST.

Ship Numbers 157 – 180 was the first contract.
Ship Numbers 237 – 247 was the second contract.
Ship Numbers 122 – 131 was the third contract.
Ship Numbers 491 – 500 was the fourth contract.
Ship Numbers 531 – 599 was the fifth contract.
Ship Numbers 806 – 828 was the sixth contract.
Ship Numbers 874 – 883 was the seventh contract.
Ship Numbers 1101 – 1110 was the final contract.

In 1945 as the last of the Landing Ship Tanks were being completed, 18 barges were under construction. The barges were similar in width and length as the LST; the keel laid on the same Ways.

During the final months of the shipyards operation a government order to construct 5 U.S. Navy YF boats, (Freight and Ammunition Lighter Ships) was being processed. From the appearance of the first YF ship under construction, its design was that of a luxury yacht rather than a war vessel.

On the following pages are recorded the history of the first 24 LSTs built at the Evansville Shipyards, compiled from the official record.

LST-157 Earned 3 Battle Stars

LST-157 was the first LST to be built at the Evansville, Indiana, Shipyard. The keel was laid down on 25 June by the Missouri Valley Bridge & Iron Co.; launched on 31 October 1942; sponsored by Mrs. DeWitt C. Redgrave, Jr.; and commissioned on 10 February 1943.

During World War II, LST-157 served in the European theater and participated in the following operations: Sicilian occupation in July 1943; Salerno landings during September 1943; and the invasion of Normandy, June 1944.

Upon her return to the United States, LST-157 was decommissioned on 9 December 1944 and was transferred to the United Kingdom under the lend-lease agreement.

She was returned to the custody of the United States Navy on 11 April 1946 and struck from the Navy list on 5 June 1946. On 5 December 1947, she was sold to Bosey, Philippines.

LST-157 earned three battle stars for World War II service.

LST-158 Earned 2 Battle Stars

LST-158 was the second LST to be built at the Evansville, Indiana Shipyard. The keel was laid down by the Missouri Valley Bridge & Iron Co., on 11 July 1942, by the Missouri Valley Bridge & Iron Co.; launched on 16 November 1942: sponsored by Mrs. H. S. Tullock; commissioned on 10 February 1943.

During World War II, LST-158 served in the European theater and participated in the following operations: Sicilian occupation and North African occupation an Tunisian operations during May, June and July 1943.

As a result of enemy aircraft action LST-158 sunk off Licata, Sicily, on 11 July 1943. She was stuck from the Navy list on 28 July 1943.

LST-158 earned two battle stars for World War ll service.

LST-159 Transferred: United Kingdom

LST-159, the 3rd., keel was laid down on 19 July 1942 at Evansville, Ind., by the Missouri Valley Bridge & Iron Co.; launched on 21 November 1942; sponsored by Mrs. Adrian G. Wrucke; commissioned on 13 February 1943. LST-159 never saw active service with the United States Navy. She was transferred to the United Kingdom on 3 March 1943 and returned to the United States Navy custody on 23 April 1946. She was struck from the Navy list on 19 June 1946. On 27 April 1948, LST-159 was sold to the Newport News Shipbuilding & Drydock Co., of Newport News, Va., for conversion to merchant service.

LST-160 Transferred: United Kingdom

LST-160, the 4th., keel was laid down on 21 July 1942 at Evansville, Ind., by the Missouri Valley Bridge & Iron Co., Launched on 30 November 1942; sponsored by Mrs. Basil T. Kehoe and was commissioned on 18 February 1943. The tank landing ship never saw active service with the United States Navy. She was decommissioned on the 4 March 1943. Transferred to the United Kingdom on 6 March 1943. The ship was returned to the United States Navy custody on 1 June 1946 and struck from the Navy list on 3 July 1946. She was sold on 5 December 1947 to Bosey, Philippines.

LST-161 Transferred: United Kingdom

LST-161 the 5th., keel was laid down on 24 July 1942 at Evansville, Ind., by the Missouri Valley Bridge & Iron Co.; launched on 7 December 1942; sponsored by Mrs. Ashley Fancy; and commissioned on 28 February 1943. The tank landing ship never saw active service with the United States Navy. She was decommissioned on 14 March 1943 and transferred to the United Kingdom on 15 March 1943. The ship was returned to the United States Navy custody on 5 January 1946 and struck from the Navy list on 20 March 1946. On 7 May 1948, she was sold to the Northern Metals Co., of Philadelphia, Pa., for scrapping.

LST-162 Transferred: United Kingdom

LST-162 the 6th., keel laid down on 24 1942 at Evansville, Ind., by the Missouri Valley Bridge & Iron Co.; launched on 3 February 1943; sponsored by Miss Marian Shapiro; commissioned on 24 March 1943. The Landing Ship Tank never saw active service with the United States Navy. She was decommissioned on 22 March 1943 and transferred to the United Kingdom. The ship was returned to the United States Navy custody on 1 February 1946 and struck from the Navy list on 16 June 1946. She was sold to Luria Brothers and Co., Inc., of Philadelphia, Pa., on 9 October 1947 for scrapping.

LST-163 Transferred: United Kingdom

LST-163 the 7th., keel was laid down 10 August 1942 at Evansville, Ind., by the Missouri Valley Bridge & Iron Co., launched on 4 February 1943; sponsored by Mrs. Monty H. Freeland; and commissioned on 24 March 1943. The tank landing ship never saw active service with the United States Navy. She was decommissioned on 29 March 1943 and transferred to the United Kingdom. The ship was returned to the United States Navy custody on 29 November 1946 and was sold to Tung Hwa Trding Co., of Singapore, on 29 July 1947 and converted to merchant service. LST-163 was struck from the Navy list on 1 August 1947.

LST-164 Transferred: United Kingdom

LST-164 the 8th., keel was laid down on 13 August 1942 at Evansville, Ind., by the Missouri Valley Bridge & Iron Co.; launched on 5 February 1943; sponsored by Mrs. Lloyd E. Miller and commissioned on 24 March 1943. The tank landing ship never saw service with the United States Navy. The ship was decommissioned on 5 April 1943 and transferred to the United Kingdom. She was returned to United States Navy custody on 29 November 1946 and struck from the Navy list on 1 August 1947. On 7 October 1947, she was sold to Tung Hwa Trading Co., of Singapore and concerted to merchant service.

LST-165 Transferred: United Kingdom

LST-165 the 9th., keel was laid down on 7 September 1942 by the Missouri Bridge & Iron Co., at Evansville Indiana, It was launched on 2 February 1943; Sponsored by Mrs. Frank Harrison. Commissioned on 3 April 1943, the tank landing ship never saw service with the United States Navy. She was decommissioned on 6 April and transferred to the United Kingdom. She was returned to the United States Navy custody on 20 March 1946. She was struck from the Navy list on 5 June 1946. LST-162 was sold to the Northern Metals Co., of Philadelphia, Pa. for scrapping.

LST-166 Earned 3 Battle Stars

LST-166 was the 10th., keel was laid down on 7 September 1942 at Evansville, Indiana by the Missouri Bridge & Iron Co.; launched on 1 February 1943; sponsored by Mrs. Vivian Massey. Commissioned on 22 April 1943, Lt. F. B. Bradley, USCG, was in command. During WWII, LST-166 was assigned to the Asiatic-Pacific theater and participated in the following operations: Occupation and defense of Cape Torokina, December 1943; Marianas operation: Capture and occupation of Saipan in June 1944; Okinawa Gunto operation: Assault & occupation of Okinawa Gunto-April 1945.

Following the war, LST-166 performed occupation duty in the Far East until the fall of

1945. She returned to the United States and was decommissioned on 3 May 1946. She was struck from the Navy list on 19 June 1946 and was sold to Dulien Steel Products, Inc., Settle, Wash., November 3, 1947 for scrapping. LST-166 earned three battle stars and the Navy Unit Commendation for WWII.

LST-167 Earned One Battle Star

LST-167, the 11th., keel was laid down on 19 September 1942 at Evansville, Indiana, by the Missouri Valley Bridge & Iron Co.; launched on 25 February 1943; Sponsored by Mrs. Harry J. Trainor and commissioned on 27 April 1943, Lt. Edward C. Simons, USCG, in command. During World War II, LST-167 was assigned to the Asiatic-Pacific theater and participated in the occupation of Vella Lavella, in September 1943.

She was severely damaged by a Japanese air attack at the Solomon Islands on 25 September 1943. Declared beyond economical repair, the ship was struck from the Navy list on 6 December 1943. LST-167 earned one award of the Navy Unit Commendation and one battle star for World War II service.

LST-168 Earned 8 Battle Stars

LST-168 the 12th., keel was laid down on 26 September 1942 at Evansville, Indiana, by the Missouri Valley Bridge & Iron Co.; launched on 25 February 1943; sponsored by Mrs. Merle Batey. It was commissioned on 3 May 1943. During World War II, LST-168 was assigned to the Asiatic-Pacific theater and participated in the following operations: Eastern New Guinea operation: Finschhafen occupation September 1943, Saidor occupation in January 1944. Bismarck-Archiplago operation: Cape Gloucester, New Britain in December 1943, also January and February 1944. Admiralty Islands landings-March 1944; and Hollandia operation in April 1944; in the Western New Guinea operations and Morotai landings in September 1944, Leyte landings-October and November 1944; Lingayen Gulf landings in January 1945; Mindanao Island landings in April 1945; Borneo operations followed with Balikpapan operation during July 1945.

Following the war, LST-168 performed occupation duty in the Far East in September and October 1945. She returned to United States and was decommissioned on 14 March 1946. She was struck from the Navy list on 12 April 1946. On 9 April 1948 she was sold to the Bethlehem Steel Co. of Bethlehem, Pa., for scrapping. LST-168 earned eight battle stars for World War II service.

LST-169 Earned 3 Battle Stars

LST-169 the 13th., keel was laid down on 1 October 1942 at Evansville, Indiana by the Missouri Valley Bridge & Iron Co.; launched on 26 February 1943 and was sponsored by Mrs. L. S. Oakes. It was commissioned on 22 May 1943. During World War II, LST-169 was assigned to the Asiatic-Pacific theater and participated in the following operations: Gilbert Islands operation-November and December 1943; Marianas operation in the capture and occupation of Saipan in June and July 1944; Leyte landing-October 1944. Following the war, LST-169 performed occupation duty in the Far East until early December 1945. She returned to the United States and was decommissioned on 12 April 1946. She was struck from the Navy list on 19 June 1946 and sold to Luria Steel & Trading Corp., New York, N.Y. on 24 October 1947 for scrapping. LST-169 earned three battle stars for World War II service.

LST-170 Earned 7 Battle Stars

LST-170 the 14th., keel was laid down on 9 October 1942 at Evansville, Indiana, by the Missouri Valley Bridge & Iron Co.; launched on 27 February 1943.; sponsored by Mrs. A. F. Sweet; and commissioned on 31 May 1943. During WWII, LST-170 was assigned to the Asiatic-Pacific theater and participated in the following operations: Bismarck Archipelago operation: Eastern New Guinea operation: Saidor occupation-January and February 1944; Admiralty Islands landing-March 1944; Hollandia operation-April and March 1944; Western New Guinea operations: Toem-Wakde-Sarmi area operation in May 1944, also Biak Island operation in May and June 1944, Cape-Sansapor operation-July and August 1944, Morotai landing-September 1944; Leyte landings in October and November 1944; Luzon operations: Mindoro landings-December 1944, Lingayen Gulf landings-January 1945; Consolidation of the southern Philippines; Mindanao Island landings-April 1945. Following the war, LST170 was decommissioned on 6 April 1946 and struck from the Navy list on 3 July 1946. On 5 November 1947, she was sold to the Biloxi Boat Wrecking Co., of Biloxi, Miss., for scrapping. LST-170 earned seven battle stars for World War II service.

LST-171 Earned 7 Battle Stars

LST-171 the 15th., keel was laid down on 20 October 1942 at Evansville, Indiana, by the Missouri Valley Bridge & Iron Co.; launched on 28 February 1943; sponsored by Miss Martha J. Miller. LST-171 was commissioned 5 June 1943. During World War II, LST-171 served in the Asiatic-Pacific theater and took part in the following operations: Bismarck Archipelago, Cape Gloucester and New Britain operation in December 1943 and January, February 1944; Also in the Saidor occupation in January and February 1944. Admiralty Islands landings in February and March 1944; Eastern New Guinea and Biak-Island operation in May and June; Cape Sansapor operation in July and August 1944; Morotai landings-September 1944; Leyte landings-October and November 1944; Visayan Islands landings-March and April 1944; Borneo operations: Tarakan Island operation in April and May 1944, Balikpapan operation in June and July 1945. Following the war, LST-171 performed occupation duty in the Far East until late October 1945 when she returned to the United States and was decommissioned on 21 May 1946. She was struck from the Navy list on 3 July 1946. She was sold on 26 September 1947 for scrapping, to Boston Metals Co., of Baltimore, Md. LST-171 earned seven battle stars for World War II service.

LST-172 Earned no Battle Stars

LST-172 the 16th., keel was laid down on 24 December 1942 at Evansville, Indiana, by the Missouri Valley Bridge & Iron Co.; launched on 12 May 1943; sponsored by Mrs. R. H. G. Mathews; was commissioned on 11 June 1943. During World War II, LST-172 performed service in China until late May 1946. Upon her return to the United States, she was decommissioned on 8 June 1946. She was struck from the Navy list on 9 June 1946. LST-172 was sold on 5 Nov. 1947 to Bosey, Philippines and converted for merchant service.

LST-173 Earned 2 Battle Stars

LST-173 the 17th., keel was laid down on 24 December 1942 at Evansville, Indiana by the Missouri Valley Bridge & Iron Co.; launched on 24 April 1943. Sponsored by Mrs. John McCone and commissioned on 18 June 1943. During WWII, LST-173 served in the European theater and participated in the movement of Convoy UGS-36 in April 1944 and the invasion of southern France in August and September 1944. She was decommissioned and transferred to the United Kingdom on 24 December 1944 and returned to the United States Navy custody on 23 April 1946. On 19 June 1946, LST-173 was struck from the Navy list. The ship was sold on 22 October 1947 to Luria Brothers & Co., of Philadelphia, Pa., for scrapping. LST-173 earned two battle stars.

S.I.G.&E.CO

LST-174 Earned 2 Battle Stars

LST-174 the 18th., keel was laid down on 1 January 1943 at Evansville, Indiana, by the Missouri Valley Bridge & Iron Co.; launched on 21 April 1943; sponsored by Mrs. Elizabeth F. Maynard and was commissioned 15 June 1943. During World War II LST-174 was assigned to the European theater and participated in the movement of Convoy UGS-36 in April 1944 and the invasion of southern France in August 1944. She was decommissioned on 21 December 1945 and struck from the Navy list on 21 January 1946. On 30 January 1947, LST-174 was sold to Ming-Sung Industrial Co., of Shanghai, China, and converted for merchant service. LST-174 earned two battle stars for World War II service.

LST-175 Earned One Battle Star

LST-175 the 19th., keel was laid down on 6 January 1943 at Evansville, Indiana, by the Missouri Valley Bridge & Iron Co., It was launched on 18 April 1943.; sponsored by Mrs. John Kuhn; and commissioned on 19 May 1943. During World War II, LST-175 served in the European theater and participated in the invasion of Normandy in June 1944. She was decommissioned on 1 March 1946 and struck from the Navy list on 8 May 1946. On 11 December 1947, she was sold to the Southern Shipwrecking Co., of New Orleans, La., for scrapping. LST-175 earned one battle star for World War II service.

LST-176 Earned One Battle Star

LST-176 the 20th., keel was laid down on 18 January 1943 at the Evansville Shipyard by the Missouri Valley Bridge & Iron Co.; launched on 15 April 1943; sponsored by Mrs. L. J. Prues; and commissioned on 12 May 1943, Lt. J. S. Salt, USCGR, in command. During World War II, LST-176 was assigned to the European theater and participated in the invasion of Normandy in June 1944. Following the war, LST-176 performed occupation duty in the Far East until early January 1946. She was decommissioned on 6 January 1946 and transferred to the Shipping Control Authority, Japan, and operated with a Japanese civilian crew. On 31 March 1952, she was redelivered to the United States Navy and served with the Military Sea Transportation Service as USNS *T-LST-176* until she was stricken from the Navy list on 1 November 1973. LST-176 earned one battle star for World War II service.

LST-177 Earned 2 Battle Stars

LST-177 the 21st., keel was laid down on 5 February 1943 at the Evansville Shipyard by the Missouri Valley Bridge & Iron Co.; launched on 16 May 1943; sponsored by Mrs. James Gibson; and commissioned on 22 June 1943. During World War II, LST-177 was assigned to the European Theater and participated in the movement of UGS-36 in April 1944 and the invasion of southern France in August and September 1944. On 11 February 1946, the ship was decommissioned and struck from the NAVY list on 12 April 1946. On 13 March 1947 she was transferred to France as a sale. She served in the French Navy as *Laita (L-9001)* until she was decommissioned in January 1962 and converted to a port depot ship. LST-177 earned two battle stars for World War II service.

LST-178 Earned 2 Battle Stars

LST-178 the 22nd., keel was laid down on 6 February 1943at Evansville, Indiana, by the Missouri Valley Bridge & Iron Co.,; launched on 23 May 1943; sponsored by Mrs. Charles Haglin; and commissioned on 21 June 1943. During WWII, LST-178 was assigned to the European theater and participated in the movement of Convoy UGS-36 in April 1944 and the invasion of southern France in August and September 194LST-178 was decommissioned and transferred to the United

Kingdom on 24 December 1944 and returned to the United States Navy custody on 12 December 1946. In November 1946, she was sold to the Egyptian Navy and was struck from the Navy list on 22 January 1947. LST-178 earned two battle stars for WWII service.

LST-179 Earned One Battle Star

LST-179 the 23rd., keel was laid down on 7 Feb. 1943 at Evansville, Indiana, by the Missouri Valley Bridge & Iron Co.; launched on 30 May 1943; sponsored by Mrs. Milford M Miler; and commissioned on 3 July 1943. During World War II, LST-179 was assigned to the Asiatic-Pacific theater and participated in the Gilbert Islands operation during November and early December 1943. The ship was gutted by fire and sunk at Pearl Harbor, Hawaii, on 21 May 1944 and was struck from the Navy list on 18 July 1944. In November 1945, she was subsequently raised, stripped, and sunk by torpedo fire in an ordnance test. LST-179 earned one battle star for World War service.

LST-180 Transferred: United Kingdom

LST-180 the 24th., keel was laid down on 8 February 1943 at Evansville, Indiana, by the Missouri Valley Bridge & Iron Co.; launched on 3 June 1943; sponsored by Mrs. Samuel M. Galls; and commissioned on 29 June 1943. On 10 July 1943, LST-180 was decommissioned and transferred to the United Kingdom. She was returned to the United States Navy custody on 17 December 1945 and was struck from the Navy list on 21 January 1946. LST-180 was sold to the Ships & Power Equipment Co., of Barber, N.J., on 10 March 1948, and was converted for merchant service.

This completes the history and listing of the first government order of 24 Landing Ship Tanks (LSTs) that were built at the Evansville Shipyard, Evansville, Indiana as chronicled in the Dictionary of American Naval Fighting Ships from the United States Navy History Internet website.

THANKS TO
THOMAS R. LONNBERG,
CURATOR OF HISTORY,
EVANSVILLE MUSEUM OF
ARTS, HISTORY & SCIENCE
providing

UNITED STATES

NAVY PICTURES

Landing Ship Tanks during construction at the Evansville Navy Shipyards and for the war zone pictures of LSTs making beach landings. 1942 ~~ 1945

Wood Foundations For Keel to be Laid on.

As the shipyard work force gained experience the construction time for a LST to be completed from date keel was laid, to commission by the Navy was drastically reduced. The first, LST-157, took over seven (7) months to be built. One year later LST-122 was completed in three (3) months, total 92 days.

LST-1110, the 167th., keel was laid on 28 Dec. 1944 and commissioned 7 March 1945, seventy (70) days later. (Note: Following 4 keels laid in 5 days.)

LST-122 Earned 4 Battle Stars

LST-122, the 36th., keel was laid down on 4 June 1943 at Evansville, Indiana, by the Missouri Valley Bridge & Iron Co.; launched on 9 August 1943; and commissioned on 3 September 1943.

During World War II, LST-122 assigned to the Asiatic-Pacific theater and participated in the following operations: Marshall Islands operation: Occupation of Kwajalein and Majuro Atolls in January and February 1944; Hollandia operation during April 1944. Marianas operation and in the Capture and occupation of Guam in July 1944.

Following the war, LST-122 performed occupation duty in the Far East and service in China until early May 1946.

Upon her return to the United States, LST-122was decommissioned on 4 June 1946 and struck from the Navy list on 3 July 1946. On 5 December 1947, she was sold to Bosey, Philippines. LST-122 earned four battle stars for World War II service.

**River View Of Coffer Dam Construction
Before Building the Launching Ways.**

LST-123 Earned 3 Battle Stars

LST-123, the 37th., keel was laid down on 5 June 1943 at Evansville, Indiana, by the Missouri Valley Bridge & Iron Co.; launched on 14 August 1943; sponsored by Mrs. C. B. Enlow, and commissioned 7 September 1943, Lt. Francis P. Rossiter, USNR, in command.

During World War II, LST-123 was assigned to the Asiatic-Pacific theater and participated in the Marianas and Green Islands operations. Took part in the Leyte landings in October 1944 and in the Lingayen landings in Luzon during January 1945.

Following the war, LST-123 was redesignated as hospital ship, LSTH-123 on 15 September 1945. She performed occupation duty in the Far East until mid-November 1945.

Upon her return to the United States, the ship was decommissioned on 22 March 1946 and struck from the Navy list, 1 May 1946. On 30 March 1948, she was sold to the Sun Shipbuilding & Drydock Co., of Chester, Pa., for scrapping. LST-123 earned three battle stars for World War II service.

Launching Skids At Sunset.

LST-124, Earned 3 Battle Stars

LST-124, the 38th., keel was down on 7 June 1943 at Evansville, Indiana by the Missouri Valley Bridge & Iron Co.; launched on 18 August 1943; sponsored by Mrs. Hunt Greathouse; and commissioned on 24 September 1943, with Ens. William A. Bartos in command.

During World War II, LST-124 was assigned to the Asiatic-Pacific theater and participated in the following operations: Marianas operation, capture and occupation of Saipan during June & July 1944. Also in the capture and occupation of Tinian in June 1944; Took part in the assault and occupation of Okinawa-Gunto during May and June 1945.

Upon her return to the United States, the ship was decommissioned on 26 July 1946 and struck from the Navy list on 28 August 1946. On 13 December 1947, LST-124 was sold to the Kaiser Co., Inc., of Seattle, Wash., and subsequently scrapped.

LST-124 earned three battle stars for World War II service.

THE SPARK WELDERS CREW THAT THE AUTHOR WORKED WITH.

The Champions of the Day Shift Softball League; Summer 1943. Front row, from left: John Brown, John Hulvershorn, Howard Hubele, John Conant, John Krause.

Back row, Jack Stewart, E.O. Hutcheson, V.H. Boyd, E.D. Darnell, Wade Pruitt.

LST-125 Earned 4 Battle Stars

LST-125, the 39th., keel was laid down on 8 June 1943 at Evansville, Indiana by the Missouri Valley Bridge & Iron Co.; launched on 23 August 1943; sponsored by Mrs. W. R. Durham. It was commissioned on 29 September 1943.

During World War II, LST-125 was assigned to the Asiatic-Pacific theater and participated in the following operations: Marianas operation with the capture and occupation of Guam-July 1944: Lette landings in October 1944; Lingayen landings on Luzon in January 1945; participated in the assault and occupation of Okinawa Gunto in April 1945.

Following the war, LST-125 performed occupation duty in the Far East and service in China until early June 1946.

Upon her return to the United States, the ship was decommissioned on 10 June 1946. She was sunk as a target, by naval gunfire, on 14 August 1946, and struck from the Navy list on 25 September 1946. LST-125 earned four battle stars for World War II service.

LST 170 At Sansapor in Dutch New Guinea.

LST-170 Earned 7 Battle Stars

LST-170 the 14th., keel was laid down on 9 October 1942 at Evansville, Ind. by the Missouri Valley Bridge & Iron Co.; launched on 27 February 1943; sponsored by Mrs. A. F. Sweet; and commissioned on 31 May 1943, with Lt. T. N. Kelly, USCGR in command. During World War II, LST-170 was assigned to the Asiatic-Pacific theater and participated in the following operations: Bismarck Archipelago operation: (a) Cape Gloucester, New Britain-December 1943, (b) Admiralty Islands landings during March 1944.Eastern New Guinea operation: (a) Saidor operation in January and February 1944: Hollandia operation-April & May 1944; Western New Guinea operations: (a) Toem-Wakde-Sarmi area operations in May 1944, (b) Biak Island operation in May and June 1944, (c) Cape Sansapor operation-July and August 1944, (d) Morotai landings in September 1944; Leyte landings-October and November 1944; Luzon operations: (a) Mindoro landings-December1944, (b) Lingayen Gulf landings in January 1945; Consolidation of southern Philippines: (a) Mindanao Island landings in April 1945. Following the war, LST-170 was decommissioned on 6 April 1946 and struck from the Navy list on 3 July 1946. On 5 November 1947, she was sold to the Biloxi Boatwrecking Co., of Biloxi, Miss., for scrapping. LST-170 earned seven battle stars for World War II service.

LST 173 Unloading in Southern France, July 5th, 1944.

LST-173 Earned 2 Battle Stars

LST-173 the 17th., keel was laid down on 24 December 1942 at Evansville, Indiana, by the Missouri Valley Bridge & Iron Co.; launched on 24 April 1943; sponsored by Mrs. John McCone; commissioned on 18 June 1943. During World War II, LST-173 served in the European theater and participated in the movement of Convoy UGS-36 in April 1944 and the invasion of southern France in August and September 1944. She was decommissioned and transferred to the United Kingdom on 24 December 1944.

LST-173 was returned to United States Navy custody on 23 April 1946. On 19 June 1946, LST-173 was struck from the Navy list. The ship was sold to Luria Brothers & Co., of Philadelphia, Pa., on 22 October 1947, for scrapping. LST-173 earned two battle stars for World War II service.

LST 245 Unloading Equipment At Tor River Area, New Guinea.

LST-245 Earned 8 Battle Stars

LST-245, the 33rd keel was laid down on 7 May 1943 at Evansville, Ind., by the Missouri Valley Bridge & Iron Co.; launched on 17 July 1943; sponsored by Mrs. Conrad L. Walker; and commissioned on 22 August 1943, Lt. Matthew J. McCabe, USNR, in command.

During World War II, LST-245 was assigned to the Asiatic-Pacific theater and participated in the following operations: (a) Saidor occupation in January and February 1944. (b) Bismarck Asrchipelago operation: (c) Cape Gloucester, New Britain in February 1944. (d) Admiralty Islands landings, March 1944, (e) Hollandia operation, April and May 1944. Also (f) Western New Guinea operations: (g) Toem-Wakde-Sarmi area operations, May 1944: (h) Biak Island operation, June 1944 and Noemfoor Island in July 1944: (i) Cape Sansapor operation in July and August 1944: (j) Morotai landings in September and Leyte landings, October and November 1944: (k) Lingayen Gulf landing January 1945: (l) Mindanao Island landings in April 1945: (m) Balikpapan operation, June and July 1945.

LST-245 was decommissioned on April 1, 1946 and struck from the Navy list on 8 May 1946. On 15 April 1948, she was transferred to the Maritime Administration for disposal. LST-245 earned eight battle stars for World War II service.

**Evansville built LST 496 sunk by enemy mines
Omaha Beach, Normandy, 11 of June 1944**

LST-496, Earned One Battle Star

LST-496, the 51st., keel was laid down on 24 August 1943 at Evansville, Indiana by the Missouri Valley Bridge & Iron Co.; launched on 22 October 1943; sponsored by Mrs. Fred McCutchan; and commissioned on 27 December, 1943; Lt. Stanley H. Koch, USNR, in command.

During World War II, LST-496 was assigned to the European theater and participated in the invasion of Normandy in June 1944.

Enemy mines off Omaha Beach, Normandy, sank LST-496 on 11 June 1944.

The Navy struck LST-496 from the ship list on 22 August 1944. LST-496 earned one battle star for World War II service.

LST-531 Torpedoed by German E-Boat

LST-531, the 56th., keel was laid down on 22 September 1943 at Evansville, Indiana, by the Missouri Valley Bridge & Iron Co.; launched on 24 November 1943; sponsored by Mrs. Marion Yoder; and commissioned on 17 January 1944, Lt. William D. Bradley, USNR, in command.

During World War II, LST-531 was sunk by a German E-boat torpedo attack in Lyme Bay, England, on 28 April 1944 while conducting a pre-invasion landing exercise. LST-531 was struck from the Navy list on 9 June 1944.

LST-544 The Mystery Ship

LST-544, the 69th., keel was laid down on 8 December 1943 at Evansville, Ind., by the Missouri Valley Bridge & Iron Co.; launched on 4 February 1944; sponsored by Mrs. Maudie M. Marlow and was commissioned on 16 March 1944. It was decommissioned on 9 August 1946 and struck from the Navy list on 25 September that same year. On 23 June 1948, the ship was sold to the Willamette Iron & Steel Co. for scrapping.

According to the History of Landing Ship Tanks built at Evansville Shipyard, 1942-1945; <u>LST-544 saw no combat service with the United States Navy</u>. Where was the ship from date of commission 16 March 1944 until date of it being decommissioned on 9 August 1946? Was it on a secret mission?

According to NavSource Online: Amphibious Archive, LST-544 was assigned to the Asiatic-Pacific Theater; received the following; Awards, Citations, and Campaign Ribbons; The American Campaign Metal, Asiatic-Pacific Champaign Medal, and World War II Victory Medal.

So where did the ship's personnel that consisted of 9 officers, 120 enlisted men, as reported by Office of Naval Intelligence-1945, spend the 23 months between the 16 March 1944 and 9 August 1946? Out of a crew of 129 men, the odds are that as of this date in 2011, approximately 33% of the crew is still living. It would be interesting to learn of their experiences during World War II service on LST-544.

LST – 555 Earned 4 Battle Stars

LST – 555 was the 80th keel, laid down on the 5 February 1944 at Evansville, Ind., by the Missouri Valley Bridge and Iron Co.; launched on 22 March 1944; sponsored by Mrs. R. E. Sharp; and commissioned on 28 April 1944. During World War II, LST – 555 was assigned to the Asiatic – Pacific theater and participated in the following operations: Capture and occupation of southern Palau Islands in September and October 1944 and was in the Leyte landings during October and November 1944. Moved on to Zambale-Subic Bay in January 1945. It took part in the assault and occupation of Okinawa Gunto in April through June 1945. The ship was badly damaged as a result of grounding off Wakayama, Japan, on 18 September 1945. Following the war, LST – 555 performed occupation duty in the Far East until early January 1946. She was decommissioned on 6 January 1946 and struck from the Navy list on 21 January 1946. Her hulk was destroyed by gunfire, 26 January 1946. LST – 555 earned four battle stars for World War II service.

LST-563

LST-563 was the 88th. keel. It was laid down on 4 March 1944 at Evansville, Ind., by the Missouri Valley Bridge & Iran Co.; launched on 1 May 1944; sponsored by Mrs. N. E, Senescall. LST-563 was commissioned on 20 May 1944, with Lt. J. B. Hockswender, USNR, in command. During World War II, LST-563 was grounded on Clipperton Island in the eastern Pacific, 670 miles southwest of Mexico, 21 December1944 and suffered extensive damage. After numerous unsuccessful attempts to free her, LST-563 was stripped and abandoned on 9 February 1945. She was struck from the Navy list on 23 February 1945

LST-566

LST-566 was the 91st., keel. It was laid down on 17 March 1944 at Evansville, Ind., by the Missouri Valley Bridge & Iron Co.; launched on 11 May 1944: sponsored by Mrs. George C. Martin; and commissioned on 29 May 1944. Following the war, LST-566 performed occupation duty in the Far East and saw service in China until early March 1946. The ship was decommissioned on 11 March 1946 and transferred to the Military Sea Transportation Service on 31 March 1952 where she served as USNS LST-566 until struck from the Navy list on 1 November 1973. USNS LST-566 was transferred to the Philippine Navy on 13 September 1976.

LST-577 Earned 2 Battle Stars

LST-577 was the 102nd. keel laid down on 3 May 1944 at Evansville, Ind., by the Missouri Valley Bridge & Iran Co.; launched on 16 June 1944; sponsored by Mrs. Christopher Roeder. LST-577 was commission-ed on 10 July 1944; Lt. Carl H. Stahl in command.

During World War II, LST-577 was assigned to the Asiatic-Pacific theater and participated in the Leyte landings in November 1944 and the Lingayen Gulf landings in January 1945. She was sunk by a Japanese submarine-launched torpedo off the Philippines on 11 February 1945. LST-577 was struck from the Navy list on 30 March 1945. LST-577 earned two battle stars for World War II service.

LST-594

LST-594 was the 119th., keel. It was laid down on 1 July 1944 at Evansville, Ind., by the Missouri Valley Bridge & Iron Co.; launched on 12 August 1944: sponsored by Mrs. Everett B. Wiley; and commissioned on 6 September 1944, Lt. Edgar Y. M. Henderson, USNR, in command. Following the war, LST-594 performed occupation duty in the Far East and saw service in China until February 1946. The ship was decommissioned on 21 February 1946; struck from Navy records on 5 March 1947. She was sold to the government of South Korea.

LST-819 Earned 15 Battle Stars

LST-819 the 138th., keel was laid down at the Evansville Shipyards on 12 September 1944. It was launched on 21 October 1944; sponsored by Mrs. Georgia E. Gilmore; commissioned on 14 November 1944, Lt. George W. Ryerson, USNR, in command. During World War II LST-819 was assigned to the Asiatic-Pacific theater and participated in the assault and occupation of Okinawa Gunto from April through June 1945. She returned to the United States and was decommissioned on 15 November 1946 and assigned to the Pacific Reserve Fleet. Recommissioned on 8 September 1950, LST-819 saw extensive service during the Korean War. The ship was decommissioned again on 24 June 1955. On 1 July 1955, LST-819 was redesignated Hampshire County (LST-819) (q.v.) after counties in Massachusetts and West Virginia. Once more commissioned on 9 July 1966, Hampshire County supported the Navy's efforts off South Vietnam until decommissioned for the final time on 19 December 1970. The tank landing ship was struck from the Navy list in April 1975. LST-819 received one battle star for service during World War II and four battle stars for the Korean War. Additionally, LST-819 (Hampshire County) was awarded 10 battle stars, the Presidential Unit Citation, and the Navy Unit Commendation for the Vietnam War.

LST- 821 Earned 10 Stars, 5 Awards

LST-821 the 140th., LST keel was laid down on 19 September 1944 at Evansville Ind., by the Missouri Valley Bridge and Iron Co.; launched on 27 October 1944; sponsored by Mrs. Hugh Robertson, Sr.; and was commissioned on 14 November 1944, Lt. C. J. Rudine, USNR, in command. During World War II, LST-821 was assigned to the Asiatic-Pacific theater and participated in the assault and occupation of Okinawa Gunto from April through June 1945. Following the war, she performed occupation duty in the Far East until early December 1945. LST-821 returned to the United States and was decommissioned in March 1946 and assigned to the Pacific Reserve Fleet. On 1 July 1955, LST-821 was redestinated as Harnett County (LST-821) (q.v.) after a county in North Carolina. Once more commissioned on 20 August 1966, Harnett County (LST-821) saw extensive service in the Vietnam War. After the war, Harnett County was again decommissioned in November 1969. LST-821 earned one battle star for World War II service. Additionally, Harnett County (LST-821) earned nine battle stars, two awards of the Presidential Unit Citation, and three awards of the Navy Unit Commendation for the Vietnam War.

LST-883 Earned 8 Battle Stars

LST-883 was the 157th LST keel laid down at the Evansville Shipyards. That was on 16 November 1944. It was launched on 30 December 1944; was sponsored by Mrs. L. D. McBride; commissioned on 23 January 1945, Lt. Winfield H. Cook in command. During World War II LST-883 was assigned to the Asiatic-Pacific theater and participated in the assault and occupation of Okinawa Gunto from in June 1945. She was decommissioned on 20 April 1946 and transferred to the Army. Recommissioned on 1 July 1950, LST-883 performed extensive service during the Korean War. On 1 July 1955, she was redesignated La Moure County (LST-883) (q.v.) after a county in North Dakota. Operating with the Pacific Fleet after the war, La Moure County was again decommissioned on 7 December 1959. Struck from the Navy list on 1 July 1960, the ship was sold to Zidell Explorations Corp., Portland, Oregon, on 30 November 1960. LST-883 earned one battle star for World War II service and seven battle stars for the Korean War.

LST-1101 Earned 6 Battle Stars

LST-1101 was the 158th., LST keel laid down at the Evansville Shipyards. That was on 22 November 1944. It was launched on 3 January 1945; sponsored by Mrs. James J. Tolson: and was commissioned on 26 January 1945. Lt. James M. Trotman, Jr., USNR, in command. During World War II LST-1101 was assigned to the Asiatic-Pacific theater and participated in the assault and occupation of Okinawa Gunto in June 1945. Following the war she performed occupation duty in the Far East until early November 1945. LST-1101 was decommissioned on 6 June 1946 and assigned to the Pacific Reserve Fleet. Recommissioned on 3 November 19 1950, LST-1101 performed extensive service during the Korean War. She was redesignated as Saline County (LST-1101) (q.v.) after counties in five states of the United States. She continued to operate with the Pacific Fleet until she was again decommissioned on 9 March 1960. Stuck from the Navy list on 1 November 1960, Saline County was transferred to the German Navy where she was converted to a minelayer and served as (Bottrop N121) until September 1971. In December 1972, she was sold to Turkey where she served in the Turkish Navy as (Bayraktar A-581) LST – 1101 earned one battle star for World War II service and five battle stars for Korean War service.

"Land-Based" LST 325 in Normandy, left high and dry on a beach by receding tide, along with other LSTs, find their anchors temporary useless. Unloading operations continue despite the lack of water.

"Land-Based" LST – 325 in Normandy

left high and dry on a beach by the receding tide, These Navy LSTs find their anchors temporarily useless, but unloading operations continue despite the lack of water.

LST-325 Heroic Memorial Ship

LST-325, a 2366-ton Landing Ship Tank was built at the Philadelphia Navy Yard; launched 27 October 1942, and commissioned 1 February 1943. On 19 March 1943, it was part of the first convoy of U.S. Navy LSTs to reach the European theater of war. LST-325 participated in the invasion of Sicily in July 1943, and in the Salerno landings in September 1943. In June 1944, the ship participated in the Normandy invasion and was stranded at low tide on 12 June 1944, as shown in the Navy Photo.

LST-325 participated in numerous operations along the northern coast of France. On 5 September 1944, she was left stranded on the beach again at Morlaix, France, after docking at high tide.

LST-325 returned to the U.S., March 1945, to Port New Orleans. It was outfitted for launching and recovering light observation airplanes. She was decommissioned July 1946. In 1951, she was recommissioned for Military Sea Transportation Service's artic operation. She was struck from the Navy list in September 1963. The ship was transferred to the Greek Navy in May 1964, serving in the Hellenic Navy and renamed it, *Syros Greek LST, (1964-2000).*

In 2000, The USS Ship Memorial, Inc., via an act of Congress, acquired permission to sail it back across the Atlantic; currently serving as a museum. It is listed on the National Register of Historic Places. Its homeport is Evansville, Indiana.

SOUP KITCHEN

TANK DECK

1/3
2/3
FULL
BACK
AHEAD
1/3
2/3
STAND
FULL
FLANK

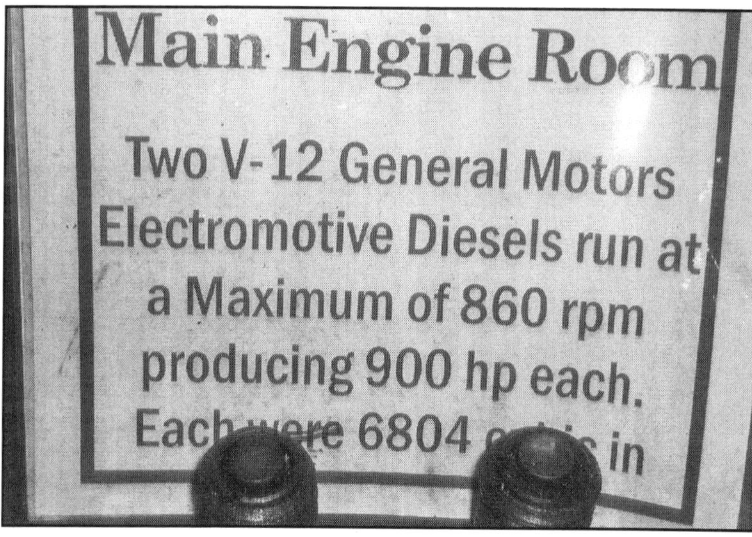
Main Engine Room
Two V-12 General Motors
Electromotive Diesels run at
a Maximum of 860 rpm
producing 900 hp each.
Each 6804 in

Chapter

26

Romance Went Asunder

I AM REMINDED OF A STORY A FELLOW WORKER TOLD ON HIMSELF AND HIS LADYFRIEND.

Bob and Sue were in their mid-forties and both divorced. She was a welder and he was a shipfitter. Bob wore a baseball cap to cover his bald head, while his beer belly lapped over his high school belt by about four inches. Sue, with graying temples and a double chin was no spring chicken.

Every afternoon at 5:20, the Wabash Cannon Ball thundered past the shipyards on its way to New Orleans. It was a modern diesel passenger train with five forest green Pullman coaches.

They thought that it would be fun to hop the train on a Monday, after their day shift, and be in New Orleans the next morning to celebrate the Mardi Gras on Tuesday, the day before Lent, and be back on the job Wednesday morning.

In their unquenchable desire to be together, like a pair of turtledoves, they had given no thought to their personal needs. They failed to bring along a change of clothes, only her big black hand bag.

The Cannon Ball Express rolled through the night as they cuddled in each other's arms in the day coach. The clicking rails lulled them to sleep, only to be interrupted occasionally by the engines blaring trumpets at every crossing.

From time to time, flickering lights from villages showed through the coach windows as the train sped south towards New Orleans.

As dawn approached, they awakened to witness the golden rays of the morning sun as it reflected from farm roofs and sparkling dew drenched the green meadows.

They did their morning exercises standing in the aisle and stretched to relieve the pains caused by the cramped quarters during the all-night ride.

The conductor passed out menus and they each ordered a light breakfast. Soon afterwards, they heard the announcement that they were nearing Memphis, Tennessee, instead of New Orleans as they expected. What a disappointment! It turned out to be "The Night Train to Memphis."

They had been badly misinformed or hadn't taken into consideration that the distance from Evansville to New Orleans was well over 700 miles. There was no way that the Cannon Ball Express could possibly make it overnight and stay on the rails.

Soon after arriving in Memphis, their nostrils inhaled that Mississippi River fishy odor, and they were ready to catch the next train back north towards Evansville. Their romance cooled as their body aromas warmed.

Time was of the essence as they had planned on taking only one day off from their jobs at the shipyards.

The next northbound train turned out to be a slow freight. They were offered a ride in the little red caboose. It was rough and noisy with no frills.

The train made many stops and did a lot of whistle blowing as it took sidings to let other trains pass or when a boxcar or occasional tanker needed to be dropped off or others picked up. The return trip seemed like it would never end with all of the loud bumping, clanging, and jerking as the couplings were released or locked together.

For twenty hours they sat on hard worn out seats, with dust and diesel fumes drifting in and out as the caboose jangled along at the end of the train.

It was impossible to carry on a conversation or get any sleep because of the noise and confusion with the brakeman, and conductor shouting orders.

They were both dog-tired and filthy dirty when they arrived back in Evansville by daybreak on Wednesday morning with only two hours to spare before reporting for work.

Their only nourishment had been a few bagels and strong black coffee that produced indigestion. Neither was worth a hoot as far as work was concerned the rest of the week.

If there is a moral to this story, it must be that love is blind and ignorance is bliss. Bob was like a slippery elm, however Sue was the clinging vine.

The Evansville shipyard vanished leaving little evidence of its existence, while the Wabash Cannon Ball continued to glide over the shiny clicking rails on its winding journey towards New Orleans.

LST-1102 Earned 1 Battle Star

LST-1102 the 159th keel, laid down on the 28 November 1944 at Evansville, Ind., by the Missouri Valley Bridge and Iron Co.; launched on 10 January 1945; sponsored by Mrs. Odette Snyder; and commissioned on 29 January 1945, Lt. L. J. Patterson, USNR, in command.

During World War II, LST-1102 was assigned to the Asiatic-Pacific theater and participated in the assault and occupation of Okinawa-Gunto in June 1945.

Following the war, she performed occupation duty in the Far East and saw service in China until early October 1947.

She was decommissioned on 21 November 1947 and assigned to the Pacific Reserve Fleet. The ship was redesigned AG-150 on 27 January 1949 and assigned the name CHIMON, after an island off the coast of Connecticut, on 1 February 1949.

On 27 December 1950, CHIMON supported operations in the Korean War.

Reclassified AKS-31 on 18 August 1951, she operated with the Pacific Fleet until 1957.

Decommissioned again on 22 April 1958, CHIMON was struck from the Navy list on 2 November 1959 and sold. To whom she was sold is not listed.

LST-1102 earned one battle star for World War II service.

Chapter

27

Even Welding Machines Needed A Sabbath

THE PUSH WAS ON TO INCREASE LST PRODUCTION AS THE WAR CONTINUED FURIOUSLY ON BOTH THE EASTERN AND WESTERN FRONTS.

The shipyards went on a seven-day week around the clock. The management even proposed that everyone work a twelve-hour shift, overlapping with the next shift in order to boost production.

That might not have been a bad idea for most of the workers such as ship fitters, electricians, shipwrights, painters, and outfitters. However, for the welders, it was a different story. There were only so many welding machines and only one welder could operate a machine at a time. The machines were kept busy so the extra welder either went to the restroom or caught a few winks of sleep.

It was time and a half pay for anyone who worked over forty hours in one week and double time pay on Sunday and holidays. Some workers enjoyed the extra moola, while others detested working on Sunday. Still others could not work the additional hours because of family problems or were unable to arrange satisfactory transportation.

For six weeks, production of LSTs increased, and twenty ships were delivered to the Navy during April and May 1944.

The 100th keel, which was LST 576, was laid on May 3rd, 1944, as dedicated workers, both men and women, achieved commendable records in building LSTs. That 100th Ship was christened on June 12th before it slid down the launching ways and moved to the dock area for the outfitters to install additional equipment.

Unfortunately, because of the increased use, many of the welding machines began to break down.

When I asked one of the maintenance men what the problem was, he replied, "The seven day continuous operation of the welding machines is causing the working parts to crystallize and break. Machines are like humans; they need to have a day of rest, a Sabbath. The machines, like Jesus said in the scriptures to his followers, 'are weary and heavy laden, and are in need of rest.' "

After six weeks of working overtime on a twelve hour shift during a seven-day work week, the company realized that the workers, as well as the welding machines, needed a day of rest, so they returned to the six day, eight hour shifts.

I didn't gain monetarily from the overtime wages. There were additional income taxes, and then I had an auto accident because of the lack of sleep. The moral of the story is, the busier we are, the more we need a day of rest, the Sabbath, just like those overworked welding machines.

Chapter

28

Fatso & His '34 Cadillac

FATSO WAS IN A MID-LIFE CRISIS WITH HIS MARRIAGE. HE HAD A ROVING EYE FOR THE OPPOSITE SEX.

His general build, excess weight, and girth, along with his spit-curled handlebar mustache and black bowtie gave him the appearance of one who might be in authority or give a person a good reason to cross over to the other side of the street in order to avoid meeting him face to face.

Before the war, Fatso operated a barbershop of his own in a small town in southern Illinois. He related that one day he was getting ready to shave a customer and had hot towels applied to the man's face when one of the usual shop loafers put down his newspaper and pulled a hand gun on the man in the barber chair, announcing that he was under arrest. The FBI agent slapped handcuffs on the fellow and stated that he was wanted for bank robbery.

Fatso told how it was a shocking surprise to him. The agent had been a customer of his for over six months and had whiled away many, many hours in the barbershop, apparently with nothing better to do. All of that time he had been on the trail of the alleged bank robber, waiting to get him in a position where he could arrest him without any difficulty.

Fatso was rooming at the same boarding house in which I stayed, owned by Mrs. Burger, located on East Columbia Street. At that time he took me for a ride in his quiet, smooth-riding jewel and offered to sell me the royal blue twelve cylinder Cadillac opera coupe for four hundred dollars. The seats were upholstered in blue mohair with all matching interior. It was a classic car with a long hood and dual wire-wheel mounts in the front fenders. It would be worth thirty thousand or more today in perfect condition as it was in 1942.

Evidently, there were not very many of that model manufactured in 1934, making it a very rare vehicle, with the 150 hp V-twelve-cylinder motor.

Fatso was one of the guards at the Evansville Shipyards. He had an infected fingernail, so he went to the first aid station. A redheaded spinster nurse treated his finger and, while talking, they discovered that they had attended the same grade school. Before they realized it, their attraction for each other had evolved into a mutual admiration society.

Fatso related how he told his wife, who was at home. "Because gas is rationed, I can't afford to drive the 180 miles round trip home but once a month. The twelve cylinder Cadillac only gets six to eight miles to the gallon."

In the meantime, he was enjoying his new arrangement with his recent acquaintance, the red headed nurse. Fatso called it a Scientific Study of Economics, whereby they each saved twenty dollars a month by sharing a two-bedroom apartment.

About a month later Fatso's brother-in-law,

Joe Byrd, stopped at Ma Burgers. He was looking for Fatso. Ma told him that Fatso had moved out. One of the men boarders spoke up, "He's living over on Virginia Ave. I'll be glad to point it out to you."

Fatso's ever-loving petite nurse, the redhead, answered the door. "No, Fatso's not here, he's at work. What do you want to see him about?"

Joe, squinting his eyes sized up the situation immediately. "I'm his brother-in-law. Fatso hasn't been home for over two months and his wife, my sister, she's having an awful hard time trying to care for their five kids and keep a roof over their heads."

A resilient old bird was Fatso. His kind may be down on the mat, but they never get counted out. The fuse blew out on his little love affair when the redhead handed him his pajamas and hat while yelling at him, "You lied to me, you no good cheater! Telling me you were single! I never want to see your smiling, good for nothing face again, EVER!"

Fatso fired up his Caddie and left as merry and bright as ever, twisting his mustache as he looked in the rear view mirror, adding a bit more spit to the twist.

I didn't feel sorry for him. He had it coming. He asked Mrs. Burger for his old room back at the boarding house, however it was no longer available because some one else had rented it.

He told me, "Had you bought my Cadillac, I could have purchased a more economical vehicle so I could have spent more time with my family."

Fatso's kind always finds someone to put the bee on! He tried to make me his scapegoat.

LST-1103 Earned no Battle Star

LST-1103 the 160th keel, laid down on the 28 November 1944 at Evansville, Ind., by the Missouri Valley Bridge and Iron Co.; launched on 13 January 1945; sponsored by Mrs. E. S. Paschall; and commissioned on 31 January 1945, Lt. Comdr. R. D. Foster, USNR, in command.

Following the war, LST-1103 performed occupation duty in the Far East until early March 1946.

She was decommissioned on 18 June 1946 and struck from the Navy list on 23 June 1947.

On 17 June 1948, the ship was sold to the Oil Transport Co., New Orleans, La. for non-self-propelled operation.

LST-1103 did not receive a battle star for World War service.

FACSIMILE OF:
FATSO'S 1934 CADILLAC LASALLE

Chapter

29

A.W.O.L. and MARRIED

THE THIRD FULL WEEK IN JUNE 1944, I DID NOT REPORT TO THE GRAVEYARD SHIFT AT THE SHIPYARDS AS USUAL.

I had been married on the previous Saturday, June 17, at the St Peter's Lutheran Church's parsonage in Stendal, Indiana. The wedding had been planned for over three months and Edna, my bride, had asked for a one-week leave of absence from her job at the Servel Corporation.

I was so much in love that my brain had taken a leave of absence and my body was floating on cloud nine and a half. I had completely forgotten to ask for a leave of absence or tell my foreman of my planned wedding.

Well, everyone, including myself, arrived in time as planned for the wedding. The Rev. H. J. Minneman performed the ceremony on the front porch of the parsonage.

Irma Marshall, my mother's niece, and her husband, Albert "Buck", drove my parents, Albert and Alice Williams, in Dad's '38 Chevrolet to Stendal, so they could be there when their youngest son got hitched. They, and Edna's mother, Martha Satkamp, witnessed the wedding ceremony.

Noble Satkamp, the bride's brother, stood in as my best man. Edna's best friend, Genevieve Lynn, was her matron of honor.

There were no pictures and no flowers. Our family and friends were all the witness we had to the shindig that afternoon.

I was so shook up, as Elvis sang later, that my mind was A.W.O.L. I never gave it a one-minute thought about paying the preacher. I found out a year or so later that Edna gave him $5.00 for his services. I never lived that mistake down. My best friend and life long mate occasionally would recall our wedding day and would remind me that it was she who had to pay the minister when we got married. Then I would remind her that it was I who paid for the license so we could be legally attached.

Later that evening, and on into the night, the Satkamp's friends, neighbors, and relatives threw an old fashion German shivaree for us by blowing horns, ringing bells, and beating on kettles.

That's when I made another mistake and forgot to pay for the refreshments, so it was my newly acquired father-in-law, August, who anted up and covered the damages for the night.

Please understand, I had never gotten married before so I didn't know what was expected of me. If someone hadn't yelled, "Kiss the bride!" I probably wouldn't have. At that moment, if someone had asked me what my name was, I most likely would have hesitated, before answering.

I took off from the shipyards the following week as we honeymooned around Spring Mill State

Park by ourselves, trying out the swings and acting like kids on the seesaw planks. My sweetie weighed six pounds more than her new attachment, so I had to sit a foot farther out on the end of the board in order to have a perfect balance as we rode up and down in mid-air, trying to miss the mud puddles as we touched the ground on our respective ends.

The week passed quickly and it was time to return to reality. We returned to Evansville to the furnished apartment where Edna had been living.

When I went to the shipyards clock alley #2 before midnight on Sunday night, June 25, to work the Monday morning's graveyard shift, there was no time card in the rack with which to clock in. Therefore, I had to step into the clock alley office and ask the night-guard clerk if he would provide me with a time card, which he immediately did.

A Navy Lieutenant was on duty in the office that Sunday night and overheard my conversation with the clerk.

He exclaimed, "You are A.W.O.L!"

It was not Lt. Sweet and I was glad of that. I certainly didn't care to have any more confrontation with Lt. Sweet after my first encounter with him during my first week of employment at the shipyards.

The Lieutenant continued; "If you are able to work at the shipyards, you are able to be in service, over there fighting. You are a deserter. You are a yellow back. You are a friend of Hitler."

Then in a commanding tone "If it were in my power tonight, I would send you to the front line of fire. You have no respect for the flag. You do not deserve the freedom you have. We have lost 30,000 men while you were fooling around."

During the previous week, I hadn't seen a newspaper and radio stations would not report war causatives until the government released an official listing.

The progress of the war was considered a military secret, withholding knowledge from the enemy. Also, the news media had a way of not reporting the losses until they could report some good news.

The clock alley clerk issued me a new time card and I clocked in. It was Monday morning, June 26th.

My welding foreman was glad to see me and asked no questions. I continued working with the skeg crew, welding in drive shaft tubes on the LSTs.

Chapter

30

My First Cooking Error

or Never Boil One Cup of Rice!

EDNA AND I RENTED AN UPSTAIRS APPARTMENT FROM EDDIE AND GENEVIEVE LYNN. THEY LIVED IN THE DOWNSTAIRS APPARTMENT.

One morning I came home hungry and tired after working the graveyard shift at the shipyards.

I was there by myself. Edna was working days at Servel, and so I had to fix my own breakfast. I dumped a cup of rice into a half-gallon aluminum kettle with a quart and half of water and set it on the apartment gas range with the burner turned on medium.

I was so tired that I laid my head on my folded arms on the edge of the kitchen table and dozed off to sleep. When I awoke, the entire apartment was filled with blue smoke and a large portion of the floor was covered with white rice. The aluminum kettle was empty and ruined, having turned black.

If the lady living across the street, seeing the smoke coming out of our upstairs apartment, had called the fire department, the rest of the story might have been something like the following that was so vivid, it seemed to have actually happened.

One fireman ran up the outside stairs with a fire extinguisher and forced open the screen door with such a jerk that the hook was never found.

His first step into the kitchen landed in the spilled rice, and he did a 180 degree twist flip, his head hitting the extinguisher as he went down on the rice covered floor, knocking himself out.

Two other husky firemen, (*ignoring the actual posted warning by the city building inspector that a 300 pound limit was on the wooden staircase,)* **started packing a hundred pounds of fire hose up the stairs, when it gave away, dumping them onto the ground.**

I heard the noise but didn't realize what it was. I was in a state of total confusion myself when an ambulance siren wailed to a stop out front.

Immediately, the ambulance staff began pounding on the front door of the downstairs apartment. A second fire truck arrived directing them around to the back where the two injured firemen lay on the ground.

I kicked the screen door open and yelled, "There's a fireman injured up here in the kitchen! How you are going to get him out, I don't know! The only other entrance is through the downstairs apartment and it's fasten with a security lock."

Most of the smoke had dispersed by the time they threw a ladder up to the second floor.

Two firemen came climbing in yelling, "Where's the fire? Where's the fire?"

All the while, I was trying to warn them to be careful as they came in skating across the floor on the white rice.

I explained that there was no fire, and I didn't call the fire department. I was fixing my breakfast and dozed off to sleep. The rice boiled over, making the smoke and the mess on the floor.

They applied smelling salts from their first aid kit to their buddy in order to bring him around. The first thing he did was to slap his aching forehead with a hand full of wet rice. A knot had risen on his head where he fell on the fire extinguisher.

He was a mess with rice all over his uniform, his hands, and in his hair. I got a towel and tried to wipe some of the rice off his hands and face while the other guys were laughing at how he looked.

They kept asking him if he thought he was able to climb down the ladder. He wanted to know what had happened to the staircase.

Finally, they got their act together, climbed down the ladder, and loaded it back onto one of the fire trucks. I was wide-awake by that time. Thank goodness, the part about the firemen was all a nightmare, but the rice was real, leaving me with my messy kitchen floor to clean up.

I spent the rest of the morning, down on my hands and knees, scrubbing the floor, cleaning up what was to have been my breakfast.

I was completely exhausted and with only a glass of milk downed, collapsed across the bed.

Edna and Genevieve worked together at Servel. They were like sisters, the very best of friends. When they came in from work they always came in the front door and then Edna would come on upstairs. She had a hard time awakening me.

She asked, "What happened to the aluminum kettle?"

As tired as I was, I did my best to explain what had really happened. There was no fire and no damage done to the apartment.

My nightmarish dream seemed so real; I was having a problem separating it from the actual facts.

By that time, Genevieve, hearing me telling Edna about the problem that I had had that morning, with the rice boiling over onto the floor, came up the front stairs asking, "For Pete's sake what happened?"

~~~~~~~TAKE A BREAK ~~~~~~~

A restaurant in New York advertised on their menu that they could provide any dish that was ordered.

A man visiting from Australia ordered a kangaroo sandwich on toast.

After waiting quite some time, the waiter came back and said, "I'm so sorry, sir, but we are completely out of bread."

Chapter

31

Walking The Rails

HEADLINE: *THE EVANSVILLE COURIER.*

"FLESH AND BITS OF CLOTHING ARE BEING FOUND FROM THE HOWELL RAILROAD YARDS ALL THE WAY TO ST. LOUIS".

"On this cold March morning, with a few snowflakes floating around, it is especially difficult to determine if there was any evidence of foul play, or even if the victim was a human or an animal.

Local authorities are puzzled. They have called for assistance from Scotland Yard in London.

The clothing items could possibly be from high tornadic winds that plagued this area last week.

More details are pending, as soon as released by authorities, after they have located the black brain box that is supposed to be indestructible according to the Associated Press."

WOW!.....Breathlessly, I rolled over and attempted to go back to sleep after such a disturbing horrid nightmare!

"BLACK BRAIN BOX? WHAT?"....... Suddenly I sat up in bed rubbing my head. "What a confusing, frightening Orson Wells type mystery? Did my walking the railroad tracks on my way home after working the graveyard shift at the shipyards, brought on this horrible dream?"

As a child, I walked the old Monon railroad tracks several times with my Father. He was interested in getting a start of a sun-fern that grew along the tracks.

The sun-fern leaf is a pale lettuce color and is much larger than the native dark green fern. Sun-fern seeds probably had fallen off boxcars that carried the seeds from far away regions where the ferns originally grew.

Anyway, when I walked with Dad, I enjoyed seeing how far I could walk on a tee-rail before I slipped off. Dad kept warning me that I might fall and crack my head, and no one had ever been able to put Humpty-Dumpty back together again!

That warning stuck in my mind, however, it was not impressive enough to keep me from trying again and again to walk the shiny rails. Walking the rails was both difficult and dangerous when wearing leather-soled shoes. The slick soles and the bright smooth rail were not compatible. Time and time again they refused to cling to each other. Each time I slipped off, I went scurrying up the crossties to keep from falling down.

Here ten years later, I'm married and working at the Navy Shipyards in Evansville. Edna and I lived on East Broadway in the Howell Addition.

One morning after having worked the graveyard shift, I decided to take a short cut home and walk up the railroad tracks. I figured that it would save half a mile of walking and I recalled having walked the rails as a lad with my Father.

I was reliving my childhood dream, walking the railroad tee-rails, but this time wearing high-top welding boots with rubber soles and heals. Hey, it was much easier to stay on the rail, and also, it was much safer. I could walk a hundred feet along the tee-rail before losing my balance and stepping off onto the crossties between the rails.

Then I heard my Father's voice quite clearly, warning me of the consequences if I slipped and fell.

Those thoughts raced through my mind as I heard the rumbling and vibration of an approaching train behind me. There were double tracks, but which track was the train on? I stepped completely away from the tracks to the left, as the freight roared past on the outer rails.

I suddenly realized that I was actually trespassing on railroad property. I had no more right to be walking the rails than the hobos had the right to ride the rails back during the depression years.

I continued to follow the tracks without realizing where I was until I was almost inside the Howell Round House, where several engines were being repaired. Many tracks led from the two main tracks. The next question was: How do I get out of here? Stepping over so many tee rails was difficult as well as frightening.

What IF I slipped and fell in the Howell Railroad Yards, cracking Humpty Dumpty? What IF a speeding train scattered my body over the next 140 miles, all the way to St Louis?

I learned a lot from walking the rails that ended in a nightmare. I decided never to do it again.

LST - 1104 Earned 1 Battle Star

LST-1104 the 161st keel, laid down on the 1 December 1944 at Evansville, Ind., by the Missouri Valley Bridge and Iron Co.; launched on January 17, 1945; sponsored by Mrs. Walter G. Koch; and commissioned on 8 February 1945, Lt. John F. Kelly USNR, in command.

During World War II, LST–1104 was assigned to the Asiatic – Pacific theater and participated in the assault and occupation of Okinawa-Gunto in June 1945.

Following the war, she performed occupation duty in the Far East and saw service in China until early April 1946. She was decommissioned on 8 July 1946. On 28 April 1947, LST-1104 was sold to the Quarterman Corp. for operation. She was struck from the Navy list on 22 May 1947.

LST-1104 earned one battle star for World War II service.

Chapter

32

Husking Corn

WITH A ROSY FINGERED DAWN IN THE MAKING, THE GOLDEN RAYS OF THE EARLY MORNING SUN SLOWLY TURNED THE SPARKING WHITE FROST INTO DEW.

By eight o'clock the hills throughout southern Indiana were radiant with fall color, the trees along the eastern horizon seemingly bursting into flames of gold, orange and red. It was the third weekend in October 1944.

My wife, Edna, and I had driven up from Evansville the night before to her parents. Their farm home was located in the hill country of Pike County, near the Dubois County line.

After a hearty breakfast prepared by Martha, Edna's mother, my father-in-law, August, asked if I would like to go with him down near the creek to help pick corn. Although I grew up in the country, my experience of shucking corn was limited.

By the time I had changed into some old clothes, August had the horses, Bert and Bade, already hitched to the wagon, ready to go. August didn't have any problem standing up in the wagon bed while holding the reins of the team. However, I needed to grasp the wagon sideboards as it rattled and bounced over the rough uneven ground.

The red leaves of the sumac, mingled with the yellow-orange sassafras, marked the eastern bounty of the property. As we neared the cornfield, a covey of quail suddenly spooked the team of horses with the whirring sound of their wings as they took flight.

By the time I had shucked out one ear of corn ready to throw into the wagon box, August had already tossed in three ears. Using a husking peg that aided the quick removal of the husk from the ear, he was pitching ears of corn against the back board on the wagon box with the rhythmic beat of a drummer, THUMP, THUMP, THUMP and then as I tossed in my one ear, it hit a flat note 'PLUNK' like an out of tune banjo.

Along about mid-morning, August yelled, "Scorchie, hear those geese headed south? Look! There are hundreds of them."

Stenciled across the clear blue sky were two 'V' formations, one a little ahead and to the left of the other. As we watched, a change of leadership took place with the lead bird dropping back toward the rear and another one taking its place as leader.

Up and down the field we walked, picking three rows at a time. THUMP, THUMP, THUMP, 'PLUNK', THUMP, THUMP, THUMP, 'PLUNK'. Over and over and over the rhythm played until after three hours we had about thirty bushels of corn in the wagon. August thought that was enough corn picking for the day, so he headed the team back towards the barn. After the team was unhitched, watered and fed, we both headed for the house where we knew a delicious meal was waiting for us.

Chapter

33

Locked Handcuffs...No Key

The war continued to rage on both battlefronts during 1944. Gasoline rationing coupons were still being issued to motorists, according to the use of their vehicle. Most drivers were complying with the national speed limit of thirty-five miles per hour. However, one driver with a 54-inch waistline enjoyed cruising as he always had at higher speeds.

A little over a year after Fatso and his "Cadillac" moved out of Ma Burger's Boarding House, his nephew, Little Joe, came to board at Ma's. One evening as I was talking with Little Joe, I inquired about his Uncle Fatso. Immediately he let me know that he had little respect for his father's brother. He took over an hour expressing his displeasure with his uncle.

"On Memorial Day weekend," Little Joe related, "we were fishing at Golden Pond on Lake Barkley down near the Kentucky-Tennessee line.'

"It was a beautiful Sunday afternoon as we headed north out of Hopkinsville, Ky. on U.S. 41 toward Evansville, Indiana. Uncle Fatso was driving his V-12, 1934 royal blue Cadillac-LaSalle opera coupe. As he speeded up to 50 miles per hour, he glanced at his rear view mirror and saw flashing red lights closing in from the rear. I was about half asleep with my head against the door glass when I heard him yell.'

"'Hey, Little Joe, wake up! We got a 'carpet-bagger' on our tail! Open the glove compartment and get that can of shaving cream and spray a bunch around your mouth and chin and then grab those handcuffs and snap one on. Then lean over here, so I can help you with the other one.'"

"Uncle Fatso braked as he pulled the car over to the side of the highway waiting for the police to catch up with him.'

"He said to me, 'I think we can get out of this if you will act like you're crazy.'"

"A deputy sheriff stalked up to the driver side with his thumbs in his holster belt. He barked like an army sergeant, 'All right! Where's the fire?'"

"Uncle Fatso worked as a guard at the Evansville shipyards.'

"He flashed his nickel-plated star badge at the officer and blathered out, 'I'm trying to get this deranged escaped convict back to the Evansville Mental Hospital for treatment.'"

"Evidently, I had soap bubbles all around my head. I know I had shaving cream up my nose as I could hardy breathe when the young deputy saw me. I was trying to rub my nose with my handcuffed fists and that's hard to do with both arms tied.'

"The officer didn't take time to check Uncle Fatso's badge. He just got real nervous and stammered out, 'Te...te...tell you what, fol...fol... follow me, I...I...I'll take you through to the next cou...cou...cou...county!'"

Little Joe took time out to rub and pick his nose as the thoughts of the shaving cream up his nose caused it to itch. Then he continued his story.

"For the next thirty miles or so, we followed the deputy sheriff's car with its red lights and siren wailing, clearing the road ahead. I couldn't see the speedometer, but I know that at times we must have been going eighty or faster. That was when the fence posts looked like a picket fence. Uncle Fatso was having the time of his life, enjoying the high-speed chase, while I sat there foaming at the mouth and having difficulty breathing. I couldn't get the darn shaving cream out of my nose. I was furious. I couldn't have gotten out of the car if I had to without some help.'

"Soon after we got through Madisonville, the deputy pulled over with his lights still flashing and had his arm out, flagging us down. I thought Uncle Fatso was going to run over the deputy as he jumped out and ran across the road in from of us.'

"He said, 'Brother, I hope this helps you out. I need to...I need to get back as, as, as I'm out of my territorial range of authority. You be careful of that con...con...convict. He...he...he looks like a dan... dan...dangerous character to...to me. The High Sheriff will probably eat my rear out for taking you this far. However, we officers are supposed to help each other. Maybe you can get another officer to take you the rest of the way.'

"You see, he thought that Uncle Fatso was a police officer of some sort.'

"The deputy made a U-turn right there on

U.S. 41 and headed back south. He was supposed to be patrolling in Christian County. That's where he had first stopped us.'

"As we started on, I asked Uncle Fatso to get the key and take the handcuffs off of me. I had had about all I could take of his stupid, idiotic joke, as bubbles and flakes of shaving cream were still floating around in the air every time I opened my mouth.'

"Uncle Fatso just sat there laughing like a silly hyena, nearly choking between words as he drawled out, 'I don't know...where the keys are. Those cuffs...were in a box of junk...that I bought for a dollar...over a year ago...at an auction.'

"I tried my best to hit him in the head with my cuffed fists.'

"I yelled, 'I'll be a deranged convict if you don't get these damn things off of me.'"

"I tried to get him to stop at the next town and get someone to cut the cuffs off but he just kept driving on. Man was I mad. I told him off in short order.'

"Dad always said that you were the black sheep of the family and now I know why. You are the most self-centered, egotistical, lowdown, err... good for nothing. No wonder Aunt Ida, threaten to sue you for divorce, and then you went and talked her out of it. Poor soul. That's when she should have taken a baseball bat to you.'

"You told her that due to the rationing of gasoline you couldn't afford to drive home but once a month to see her and your children. Then you take me on a fishing trip to Tennessee. You're a liar and a cheat, buying black market gasoline ration coupons. This old car can't possibly get more than five miles a gallon the way you drive.'"

Ma Burger interrupted Little Joe, telling him that he was wanted on the phone.

After he came back, he sat down and asked me, "Did I tell you about Uncle Fatso buying black market gas ration coupons?"

"Yes," I replied, "You said something about he took you to Tennessee on a fishing trip but claimed that he couldn't afford to go see his wife and family in Illinois."

"Right, and getting back to the rest of my story about Uncle Fatso, I told him, he lied to that sheriff's deputy and then insisted that I to go along with his insane idea for me to act like a crazy idiot that I must had been. He thought it was funny to outsmart the law.'

"As we were coming into Henderson, I told him to hold his horses! Maybe the truck stop would have something that would cut the shackles off. And, I warned him. 'You take me across the state line into Indiana with these handcuffs on, and I'll scream that I'm being kidnapped and the Indiana authorities will believe me. It will go hard on you, uncle or no uncle. Maybe a few years in prison might do you some good.'"

"Uncle Fatso asked the attendant if he had bolt cutters that he could use to cut my handcuffs off. In offering an explanation, he said that we had been playing around with the cuffs when they got locked and that he didn't have a key to unlock them. He offered twenty dollars to anyone that would cut them off. After listening to Uncle Fatso's story the attendant didn't know what to believe.'

"Several truck drivers had gathered around, listening to Uncle Fatso's story and were skeptical. They figured that I was an escaped prisoner from a jail and the fat man was trying to help me.'

"About that time, the town marshal drove in. Leaving his police car radio popping and cracking off and on mingled with static, he walked over and asked, 'What's going on here?'"

"While Uncle Fatso was once again trying to explain, the officer kept looking at me to see if I disagreed with anything.'

"The marshal removed his hat, scratched his head for a moment, and commented, 'There's one way to find out if either of you are wanted by the law.'"

"He went over to his patrol car and radioed somewhere. Guess he was checking to see if I was an escaped convict or something. His radio kept popping on and off and after several minutes he came back into the garage and said, 'Well, neither of you are on a wanted list as far as I can determine. I have no reason to not believe your story, although it does sound a little fishy. Did I hear you say that you had lost the handcuff keys?'"

"That was when I spoke up and told him just how it was. 'My uncle told me to put the handcuffs on, then said he didn't have the key to take them off. Had I known that, I'd never put them on in the first

place. He made me do it.'"

"The marshal squinted his eyes and took a sharp look at both of us and asked me, 'Are you suggesting that there's a problem between you two? Anything you want to discuss?'"

"I saw Uncle Fatso bite his upper lip and I knew that he didn't like what I had said. Then he extended his right index finger, making a slicing motion from his left ear lobe across his throat as to say, 'I'm in trouble. You keep your big mouth shut!'

"I caught his drift but I was mad. He didn't want me to say anything. I looked him right in the face and mouthed, '*You fess up, or I'll spill the beans.***'**

"He read my lips. I'd had those cuffs on for over three hours and I was hurting.'

"Uncle Fatso picked up the bolt cutters and handed them to the town marshal. 'Here, cut his irons off.' Man...that was when I threw a fit! It sounded to me like he said to cut my ARMS off!'

"Then Uncle Fatso tried to apologize to me. He asked me to forgive him for being such a lousy, no good uncle. He also admitted that he was the black sheep of the family and promised to try to mend his ways.'

"Man, it felt so good to get those cuffs off! For several minutes I just stood there rubbing my arms and hands trying to restore circulation.'

"The trucks drivers, who had been milling around listening to the conversations, continued to discuss among themselves what had transpired. It probably will be one of those unbelievable stories that they will keep talking about for months.'

"I thanked the marshal for his help. He said, 'It's all in a day's work in my line of duty. I'm glad we got your problem solved. Your uncle appears to have a severe mental problem to have put those old handcuffs on you for no apparent reason and with no key to take them off. You better watch him from now on. No telling what he'll try to get you involved in next.'"

"As the town marshal jumped into his car to leave, Uncle Fatso ran yelling, 'Hey wait a minute.... I have a confession to make! I'm guilty of impersonating a police officer by using my shipyard guard badge, and another thing, in regards to the national speed limit of 35 miles per hour, that 12-cylinder car idles faster than that. It doesn't run smooth until it reaches 50 miles per hour. Here, I'm guilty...put your handcuffs on me.'"

"But the marshal didn't, even with Uncle Fatso's admission of guilt.'

"Guess he never had a person to admit to a real crime before and really didn't act like he knew what needed to be done in a case like that.'

"The marshal said that according to law he could hold Uncle Fatso overnight on the charge of assault and battery. He would need to appear before the local Justice of the Peace on Monday morning.'

"However, he said that he had no reason to detain me. Since I had no other transportation, I had to stay with Uncle Fatso overnight in their lockup or "hoosegow". It wasn't too bad as they brought supper and breakfast to us and told us that we could use the rest room there in the police station. However, the benches made for hard beds.'

"The following morning, the town marshal took Uncle Fatso before the Justice of the Peace to be interrogated on an assault and battery charge. That was because he had me put the handcuffs on when he knew he didn't have a key to take them off. I was their star witness of the fact. The J. P. didn't seem much interested in his confession about the black market items that he claimed to have bought.'

"They fined him $30.00 dollars, with a warning to never again impersonate a police officer. Then Uncle Fatso didn't even thank the judge for being so lenient on him. Instead he was complaining about the fact that everyone kept referring to his car as a Cadillac, when actually, according to him, it was a very rare LaSalle."

Little Joe turned to me and said; "Now you know why I don't have any respect for my Uncle Fatso. He promised to change, but did you ever see a skunk that was able to changed the color of the white stripe down its back? In my uncle's case, it was a yellow stripe."

LST-1105 Earned 1 Battle Star

LST-1105 the 162th keel, laid down on 5 December 1944 at Evansville, Ind., by the Missouri Valley Bridge and Iron Co.; launched on 20 January 1945; sponsored by Mrs. William L. Peters; and commissioned on 13 February 1945, Lt. Lionel B. King, USNR, in command.

During World War II, LST-1105 was assigned to the Asiatic-Pacific theater and participated in the assault and occupation of Okinawa-Gunto in May and June 1945

Following the war, she preformed occupation duty in the Far East until mid-February 1946.

She was decommissioned on 29 May 1946 and struck from the Navy list on 19 June that same year.

On 20 May 1948, the ship was sold to the California Co., for operation duty.

LST-1105 earned one battle star for World War II service.

Chapter

34

Kentucky Fishing Trip

WORKING AT THE SHIPYARDS WAS A GOOD EDUCATION IN SOME RESPECTS. A FEW MONTHS WERE EQUAL TO ATTENDING COLLEGE FOR A YEAR.

I'll admit that a few of the things that I learned weren't worth finding out about; for example, there was that fishing trip to the Blue Grass State.

Charlie Brown, a fellow worker, was always talking about fishing and claimed that the really big ones were down in old 'Kentuck' where he grew up.

I finally agreed to go with him on a bright Saturday morning. However, I told him that I was no fisherman. "The only time I ever went fishing was when my folks were going some place and they sent me over to stay with my cousin. We sat in the hot sun on the side of his pond for hours casting with a cotton string affix to a sassafras pole, using dead worms for bait and never caught nothing, other than a lot of chigger and mosquitoes bites."

Charlie was driving, and he knew where he wanted to go. His old 1934 Chevy pickup rattled across the Henderson Bridge and on down US Highway 41 a ways before he made a left turn onto a country gravel road and headed up into the hills.

Charlie, said, "I think I turned too soon."

"What do you mean, turned too soon? I thought you said you knew this area like the back of your hand."

"Yep, I do, but for some reason it just doesn't look natural. Think we'd better go back to the highway and go on down a little farther. So he turned around and started back to the highway. After about a mile we came to another road leading off into the hills and old Charlie stopped and seemed to be studying the location.

I asked, "What is your problem now? Are you going to take that road?"

"Well, I believe that is the one I should had taken."

So off we went down this narrow creek gravel lane with grass growing up in the middle. We bumped over a hill and down into a creek bottom that opened to a green meadow where a large herd of diary cattle grazed. Charlie seemed satisfied that this was the right road to where the big fishes could be caught.

"We'll come out over near where Green Pond and Rough River join." After an hour or so we came into Calhoun across the river from Rumsey.

Charlie said, "Now I know where we are at!"

After another hour, he stopped near Hartford at a turn in the river and declared, "This is it! This is the place. Now for my fishing gear."

Looking back in the bed of the pickup, there were his fishing rod and reel, but no bait box. Scratching his head, Charlie admitted that he had forgotten his fishing box.

I remarked, "With no bait or lures, you can't catch very many fish."

"Well, at least now, you know how to get here!" Charlie laughed.

"I doubt it. I would need a map to find my way back to Evansville. You got me lost."

Charlie shook his head, "Well, while we are here, let's see if we can locate a good spot to build a cabin. I'll ask around and find out who owns the ground and see how much they want for a lot."

After we walked up and down the roadway for quite some time, Charlie decided that just around the bend would be an ideal place as it was on higher ground with a grassy area of about half an acre.

We got back into his pickup and drove on around the bend about a quarter of a mile to the first house that we came to. It was just a weathered slab-board shack, situated on pilings four feet above the ground. A ramp led up to the front porch.

I got out with Charlie as a woman opened the screen door wanting to know what we were up to. Charlie asked her if she knew who owned the land down around the curve in the river.

She answered, "I do. Why do you want to know?"

From where I was standing, I saw a baby crawling on the floor with what looked to be a loaded diaper. A Leghorn hen sitting on the back of a chair deposited her calling card onto the floor.

I thought I heard some pigs grunting, and as I stepped closer to the edge of the road I could see under the house. A sow and several little pigs were grunting and fighting over the slop in a trough made from an old twenty-gallon gasoline tank. A pipe came down from the kitchen sink so any wastewater ran into the trough for the pigs to drink.

I was considering the situation while Charlie was talking with the woman about the piece of property he was interested in. Before long, they had made a deal for Charlie to get someone to survey off a corner down by the river bend.

I sort of interrupted and said to the woman, "That looks to me like that's not very sanitary having pigs under the house.

She snapped back, "We've never lost a pig yet."

On the way back to Evansville, Charlie said, "You shouldn't have said anything to that woman about her pigs. That's none of our business how she raises her pigs."

"Well, I wasn't thinking about the pigs, I was thinking about that poor little baby crawling around on that dirty floor with all of those pigs under the house, not to mention the chickens in the kitchen."

Charlie replied, "Don't you dare report what you saw to the county health department. She might renege on my deal to buy that lot off of her."

The last time I talked to Charlie, "I said Charlie; Remember, the time you took me down to your favorite fishing spot?"

He replied, "Sure thing, and you remember that dirty kid? Why he's our county Prosecutor and has been for several years."

Chapter

35

Nudist Across The River

A RUMOR FLOATED AROUND THAT A NUDIST COLONY WAS ACROSS THE RIVER FROM THE EVANSVILLE SHIPYARD.

The river was approximately two miles wide at that point and no activity could be seen over on the Kentucky shore with the naked eye.

Eddie Lynn, who was my landlord, had a great sense of humor. He said,

"If we had a large magnifying glass, we could watch them sunbathe."

I happened to remember that my father had a telescope that he used to view the full moon many years ago. Now, if there was some way that I could get him to let me borrow it, say for a week or two or three, then, just maybe, we could set it up on Reitz Hill, which was a mile from the river, so we could enjoy the scenery on a warm Sunday afternoon.

Well, the weather turned cold early and we never made connection that fall. However, Eddie's memory was sharp. As soon as it started getting warm the following summer, Eddie suggested,

"Get your dad's telescope back down here so we can have some fun with it. We'll do a little moon viewing across the river when the weather gets hot."

Dad had several questions when I asked to borrow his telescope again. I told him my friend Eddie wanted to take a look at the Kentucky moon along with other scenery across the river.

Evidently my answer was sufficient because Dad got the scope out of its hiding place in the front room closet and I brought it back with me to Evansville.

The hot weather finally arrived along about the middle of July, so Eddie and I could verify the rumor about the nudist colony.

On a very warm Sunday afternoon I told my wife Edna, "Eddie and I are going up on Reitz Hill."

Eddie spoke up, "We'll be back after awhile. I doubt if you girls would want to go along and look over the shipyards."

Edna replied, "Be careful and don't get into any trouble. Genevieve and I will take a walk while you guys are gone."

So off to Reitz Hill we went. We set up the telescope on the tripod and took a quick look at the shipyards before swinging it around to view across the river.

The Ohio River makes a large Horse Shoe Bend across from Evansville and the supposed nudist colony was located near White City on a sandbar down river towards Henderson, Kentucky.

That would have made it located some five or six miles down river and on the northeast side of the bend from where the shipyards were located.

Eddie exclaimed, “I see a dog!”
“Any body or bodies?” I inquired.

“Yes, several,” Eddie replied, “but I can’t tell if they are male or female, but it’s a camp alright.”

“Let me have a look. My eyesight might be better than yours. Eddie, I see one that either has on red clothes or a severe sunburn. You know, we could drive over there. It would probably take an hour or longer. We would have to find the road from Henderson out to White City. It would probably be at least twenty-five miles by car.”

Eddie answered, “Yes, but if we had a motor boat we could find out within a few minutes, if they are in their birthday suits, or if it’s a false rumor.”

I sighed a long breath of fresh air as I stood up, and stretched my back, having been bent over looking through the telescope. “Eddie, you know. ...That we may never know.”

~~~~~~TAKE A BREAK ~~~~~~

A large hole was discovered in the wall surrounding the Carefree Nudist Camp. The local police, the sheriff department, and the state police are all looking into it.

One nudist said to another nudist, “What‘s that frog croaking about? He croaked all night long.” “Oh!” The beautiful young nudist replied. “That must be the handsome prince that begged me to kiss him last night and he went blind. Guess he wants me to kiss him again to restore his sight.”
~~~~~~

LST - 1106 Earned 1 Battle Star

LST-1106 the 163rd keel, laid down on the 9 December 1944 at Evansville, Ind., by the Missouri Valley Bridge and Iron Co.; launched on 24 January 1945; no sponsor listed. She was commissioned on 16 February 1945, Lt. Jack Flinn, USNR, in command.

During World War II, LST–1106 was assigned to the Asiatic – Pacific theater and participated in the assault and occupation of Okinawa-Gunto in June 1945.

Following the war, she performed occupation duty in the Far East until late September 1945.

She was decommissioned on 2 August 1946 and struck from the Navy list on 8 October that same year.

On 13 June 1948, the ship was sold to the Walter W. Johnson Co., for scrapping.

LST-1106 earned one battle star for World War II service.

Chapter

36

Pike County Coal Bank

GLISTENING CHUNKS OF WET COAL ROLLED OUT OF THE HILLSIDE AS TWO STICKS OF DYNAMITE EXPLODED UNDER A CLEAR BLUE SKY ON A SATURDAY AFTERNOON IN LATE SEPTEMBER OF 1944.

I had worked earlier that morning at the Evansville Shipyards on the graveyard shift. After catching a few winks of sleep, my wife, Edna, and I had driven up to her parents Pike County farm in southern Indiana for a short visit. I had to be back at work by midnight.

My recently acquired father-in-law, August, asked, "Scorchie, would you like to ride on the wagon with me, down to the coal bank? You can watch while I load up the wagon."

I immediately recognized that this was a very good way for the two of us to have some quality time together. Also, it would show that I was willing to help even if it required getting my hands dirty.

With the two horses, Bert and Bade, hitched to the pole wagon, we were soon rattling down the old wagon road to the coal bank along Cup Creek.

August already had two holes drilled for the dynamite charges. After uncovering the wooden box where the two sticks of explosive were hidden, he placed them with the caps and ran the fuse wire back up the hill two hundred feet to the battery box. In a matter of seconds the dynamite explored, shooting dirt and smoke some thirty feet into the air. Large blocks of coal rolled down into the creek bed.

I immediately started to help load chunks of beautiful black carbon. I had never seen an outcropping of a coal vein. It was so amazing to me for a person to have their very own coal supply.

Before long we had over a thousand pounds of coal loaded, about as much as the team could pull back up the hill to the house. The horses strained to move the load as the wagon wheels sank into the soft earth along the creek bank.

The wagon's rattling sounds were replaced with a more solid humpy-bump, bump-thump under the weight of the load.

August let the horses rest every few hundred feet. The pull upgrade was tiring to the team.

We were soon back, at the house, where August unhitched the team and turned them out in the barn lot where they had fresh water to drink.

We lifted off the large blocks before starting to remove the wagon bed boards. August asked me to stand at the rear of the wagon while he stood at the front. With both of us lifting on the same floor pole, we made an opening. Then we removed the next pole. Each time we removed a bed pole some coal fell on the ground. Soon the wagon was empty.

We could have gone back to the coal bank for another load, however, August thought that we had worked enough for one afternoon.

Chapter

37

Lissa's Secret

STEVEN STANFORD WAS ONE OF THOSE GOOD OLD BOYS FROM THE GREAT STATE OF TENNESSEE who came up north during WWII and got a job at the Evansville Shipyards as a welder building LST boats.

When Steven received his induction greetings from Uncle Sam, he ran to the recruiting office and enlisted in the U. S. Navy.

After two years in the South Pacific, he took a R & R leave in Buenos Aires, the capitol of Argentina. While there, he met Lisa, a beautiful dark haired senorita. After several dances of the rumba and a few drinks of a homebrewed coquille, they decide to get married.

After the war, they settled in Tennessee, where they resided for some ten years. Lisa became very homesick for her native homeland. Steve agreed to arrange for them to take a vacation to Argentina so Lisa could visit her old stomping grounds and friends.

Within a few weeks, Steve was ready to return to their large cattle ranch in Tennessee. Lisa begged to stay in Buenos Aires where she had worked before the war as an escape artist in a touring rodeo show.

Her look-a-like cousin Lissa, who was two years younger than Lisa, offered to take Lisa's place as Mrs. Steven Sanford, if Steve would agree to marry her.

At first Steve protested that it wouldn't be right for her beautiful dark haired cousin to take his wife's place. However, on second thought, Steve agreed, providing his wife, Lisa, would sign certain legal papers annulling their marriage and relinquish all claims she had concerning their real estate property and her United States citizenship.

So, cousin Lissa, with the similar name, dropped the second 's' out of her given name and came to the United States as the wife of Steven Stanford. No one was any wiser that Lissa wasn't the original Lisa and Steve was happier because the new Lisa was more energetic and enthusiastic.

There was one neighbor lady who seemed to notice a slight difference in Steve's wife. Her comment was that the two-month vacation had agreed with Lisa as she looked so much younger and much happier.

~~~~~~~TAKE A BREAK ~~~~~~~

Steve: "Honey, I've got two tickets for the ballet."

Lisa: "Oh, lovely, I'll start getting ready."

Steve: "Yes, do--the tickets are for tomorrow night.
~~~~~~~

Chapter

38

All Sorts of Characters

THEY CAME FROM ALL WALKS OF LIFE FOR EMPLOYMENT AT THE SHIPYARD.

There were former coal miners, farmers, oil field workers, engineers, ex-convicts, mechanics, merchant marines, ministers, moon shiners, office employees, and yes, cowboys with their boots on. You name an occupation and no doubt someone of that trade could have been found working at the Evansville Shipyards.

I know because I encountered them by chance, while working on the job. In some cases, there was an open exchange of intimate feelings and in others a true confession of former activities that they were not proud of. Then there were those who tried to force their beliefs on others whether wanted or not.

One worker, Al, talked freely about the five years he had spent up at Michigan City making license plates. He wasn't proud of the fact, however, for him it was an education that taught him the error of his ways. He was a much better citizen, after having been in an institution that offered him help in going straight after having been released from prison. Al was 19 at the time he was sent to prison, during the depression of the early thirties.

Al was the eldest of a family of ten children. His parents feed and clothed all ten, working as tenant farmers in central Indiana. Al felt that he needed to help his parents, so he quit school when he reached the age of sixteen in his freshman year of high school and went to work for a local farmer, shucking corn and feeding cattle. The pay was $20.00 dollars a month, plus his room and board. Al gave three-fourths of his earnings to his mother to help with the other kids' school needs.

In the fall of 1932, Al's boss, the farmer, sent a fattened calf to the Indianapolis Stock Yards. After the trucking and commission fees were deducted, the farmer received a statement that he owed 34 cents to the Commission House.

The farmer told Al that since he didn't have any money to pay him, he needed to look for work elsewhere. Al started hanging around the poolroom in town, hoping to pick up a job now and then.

One day a total stranger in the poolroom asked Al if he would like to join him in a game. Al told him that he didn't have any money. The stranger asked if he would like to have a job driving him around for a few days. That was quite an offer for a young lad, especially since it was a new 1933 Ford V-8 sedan that he would be driving. The well-dressed fellow paid for Al's eats, and on a few afternoons, they drove around central Indiana visiting several small towns.

One rainy morning, he asked Al to drive him over to a certain town and told Al where to park while he transacted some business.

Al said, "I wondered what kind business he was in, however, I didn't have to wait long to find out." His dark suited boss came running, jumped into the car and yelled, "Get going. I just robbed the bank!"

"Needless to say, we got caught, but not until after we had hid the loot. That was over ten years ago. I didn't rob a bank. I had no gun. They said that I was an accessory to the crime because I was driving the getaway car."

In a husky voice, Al continued. "The other fellow died while in prison. The other Saturday, I asked my wife if she would want to take a ride with me. We drove up to a certain town and then out into the country a ways. After all these years I thought someone might have found the money."

Al continued with his story. "After driving out a good ways, I turned around and told my wife, it couldn't be this far out. There was only one culvert that we had passed so that one had to be the place. I was more nervous than I had ever been in my life, as I lifted off several large rocks, and saw the leather bag. I hid it under the car seat and didn't get it out until after dark."

After wiping perspiration from his forehead with his handkerchief, Al sat down on a crate of pipe fittings and resumed his story. "When I counted the bills, it amounted to thirty-five hundred dollars. I served five years of my life as payment for my ignorance, and the other fellow is dead. If I turned the money over to the bank authorities, they wouldn't know what to do with it. It was written off years ago. I figured that I might as well make use of the money and pay off some old bills and use the rest to buy a better pickup. Do you blame me?"

What a confession! One thing was for certain, I wasn't going to get involved in his escapade of ten years ago. I could sympathize with him, as I too might have been tricked into the crime under the circumstances he explained to me.

~~~~~~~TAKE A BREAK ~~~~~~~

After robbing the bank, the thief was so nervous, he rushed home and began to saw the legs off his bed. His wife exclaimed, "What in the world are you doing?"

"We need to lie low for a while." Whispered her husband.

~~~~~~~cwms~~~~~~

Prison Guard: "Sir, I'm sorry to report that ten prisoners have broken out with ..."

Warden: "Blow the whistles, sound the alarms, alert the police----What are you waiting for?"

Prison Guard: "Shouldn't we call the doctor first---it looks as if it might be measles.

~~~~~~~cwms~~~~~~~

Wholesome---the only word from which you can take the whole and still have some left.
~~~~~~~

Chapter

39

More Character Stories

EVANSVILLE WAS ONE OF THOSE WAR BOOM TOWNS WHERE TWO-THIRDS OF THE INDUSTRIES HAD WAR CONTRACTS.

The shipyards went nonstop making 167 LSTs plus 34 other types of boats while Republic Aviation turned out 6,242 P-47 Thunderbolt fighter planes by the end of 1945. Chrysler-Plymouth produced more than 3.2 billion rounds of .45-caliber ammunition. The Servel plant made the wings for the P-47 while Hoosier Cardinal produced the tail sections.

Workers came from every corner of the nation for employment in the Evansville area. 'Red' Ohadi, which is Idaho spelled back-wards, was part Indian from the 10,000 Lakes of Upper Minnesota. That is Paul Bunyan country and I asked 'Red' if he was related to the famous man after he told about trout fishing in the Mississippi River where it ran north instead of south.

'Red' was a pleasure to work with. I learned a lot of things just listening to him tell about his experiences while working on the railroad he called, 'The Bemidji Line.' He said, "I was a fireman on the Fargo-International Falls Railroad during the early depression years."

“Shoveling coal into the firebox of a steam locomotive all day will turn your complexion from white to black. My kids were scared of me when I came home. I was a stranger to them.”

‘Red’ continued with his story as we entered one of the ‘C’ sections to do pick up welding.

“Early one morning, late in January of 1936 as I walked to the rail yards there in Bemidji, everything was quiet except for the crunching and squeaking of the snow under foot. With every step, the sounds echoed from rooftop to rooftops until it became almost deafening. The wind that had howled endlessly throughout the night was now calm. Stars twinkled in the crisp blue sky. My every breath hung as a fog.

In the dim light shining through the station window, I looked to see what the temperature was. The three foot tall thermometer by the west door registered to sixty below. I couldn’t believe what I saw. All of the mercury was snuggled down into the small glass ball at the bottom!”

Later, after we had completed all of the welding in that section, ‘Red’ continued his tall story that he declared was true.

“Bemidji was about midway between the two terminals. On that particular morning we pulled out of the station headed for Fargo. Every time the engineer blew the whistle, steam froze into chunks of ice rattling down on top of the engine. The run of about sixty-eight miles normally took a little over two hours.”

A few days later as we were taking a break from welding, while working on a scaffold under the main deck, 'Red' tried to finish his story that he had started several days earlier.

"Like I was telling you about working on the railroad up in Minnesota. There, a few miles out from Bemidji, was a large railroad cut that the wind had filled with blown snow, like I was telling you about when it was so cold. Well, the engineer ran the engine into the drift until it stalled.

Then we backed up about one-half mile and made a run for it, trying to break through the drift. That time the engine rammed in so far that an avalanche of snow tumbled down the smoke stack and flooded the firebox. There we were, buried in a snow drift with only a shovel full of hot coals left to rekindle the firebox."

I was all ears as I asked. "What did you do then? How did you get out of there?"

'Red' continued, "That's another story that I won't go into details about. To make a long story short, it was three days later, after endless shoveling and sleepless nights, when we managed to break loose from the clutches of nature."

On another occasion 'Red' took a spell of talking and related the following story. "Every once in awhile the engineer was known to take a snort from the bottle he tried to keep hidden in an inside pocket of the blue denim jacket that he wore. One morning he didn't show up at the appointed time for work. The train schedule had to be met or we all would lose our jobs. The crew that had completed their run went off duty. After talking with the brakeman and conductor, I shoveled in the coal and took over as engineer and fireman. I had ridden many a mile with the engineer and knew every move he made." After a brief pause, 'Red' finished his story.

"We pulled the train into Fargo on time and when the Rail Superintendent found out that the engineer hadn't reported in, he gave me the job as a full time engineer. The other fellow lost his job over boozing once too often.

When they came out with the diesel engines in 1936, I wasn't qualified to operate them. I was out of a job. That was when I moved my family down here to Indiana."

~~~~~~~TAKE A BREAK ~~~~~~~

Susie running to her mother: "Mom! What was the name of the last station our train stopped at?"

Mother: "I don't know---How many times have I told you to not bother me---can't you see I'm reading?"

Susie: "Well, that's to bad, because that's where little Benny got off."

~~~~~~~cwms~~~~~~~

Keep smiling--it makes others envious as they would love to know what secret you have up your sleeve.

Chapter

40

Incognito Personalities

UNDER ASSUMED NAMES OTHER CHARACTERS WERE WILLING TO REVEAL THEIR PECULIARITIES AND ACTIVITIES.

Everyone has a story and so it was with several people I worked with at the shipyards. The stories that follow are more true stories about life and lessons to be learned.

One co-worker had farsighted plans, however, his bride-to-be wasn't in agreement.

P. D. Bullock told me that while driving around one day near the Boonville Pike on the northeast side of Evansville, he saw a real estate sign nearly hidden by ten-foot tall horseweeds.

10 ACRES FOR SALE
A GREAT INVESTMENT ONLY $3,495.00
Phone 555-THE-BANK
THE OLD NATIONAL BANK BUILDING

The following Sunday afternoon P.D. took his fiancé out to have a look at the property. He commented, "Maybe we should try to buy that piece of ground as it would probably be a good investment."

His lady friend replied, "$350.00 per acre?"

"Are you serious? That's an awful price to pay for one acre of dirt!" He agreed that it did seem a little high. *(It was in the Vicinity where Robert's Stadium is now located.)*

* * * * * * * * * *

Another fellow I'll call 'Don' related, "My mother and stepfather had great hopes for me. They wanted me to grow up and amount to something.

"I had the idea that if I could only move out to Hollywood I could get into pictures since I just knew that I could put on an act as good as Tom Mix or that Hop Along - what's his name…'Cassidy' fellow. The only difference was, they had a horse and I didn't." Don's reminiscing continued.

"My step-dad and mom pinched pennies so I could attend John Herron Art School two years in Indianapolis. I had a great future in commercial art, but I blew it. I followed the crowd. I didn't have sense enough to say no. They said, 'Do this, do that' and I just sent myself to hell. You name it; I probably did it, for the lack of common horse sense.

"Looking back, I can see it all clearly. That's the way it went. I lost my talent, lost my health, lost my job, and ended up in the psychic ward."

After a minute or so, through misty eyes, arms folded, his left hand holding his chin in a manner to support his head, with the tip of his left index finger resting on the end of his nose, Don continued.

"My dad died when I was very young. Mother remarried a good man, but he was strict, and had his way of doing things. Mother never crossed him. She was such a good person.

"After I got straightened out, I bought a place here in Evansville so I could be of some help. Mother's gone now. I really miss her. She was always proud of me, even though I didn't turn out as I should have."

* * * * * * * * * *

Then there were Robin and Richard, better known as Rob and Dick, who were motorcycle enthusiasts. They shared a story with me that sounded like a tall tale, but evidently it was actually true, about taking a trip a few years back in 1936 on their new Harley Davidson motorcycles.

Dick said, "We had planned for quite some time to tour the Rocky Mountain States on our new bikes. The factory where we worked always shut down for two weeks during the summer, so we figured that's as good a time as any for a trip west."

Rob piped up, "Yes, we both lived all of our lives here in Evansville, and we figured that it was high time that we got out and saw how other people lived. In order to get a head start, we took off real early, before daylight on a Saturday morning."

Dick continued, "Riding up to Princeton, we headed west and crossed the Wabash River at Mt. Carmel, Illinois. We encountered fog in a few low lying places. We stayed on Illinois 15 across the state into St. Louis where we took US 40 west."

Rob interrupted, "Having watched dozens of western movies, we both thought we knew what to expect, but it was nothing like the movies. The invigorating aroma of new mown hay was so refreshing. We had never experienced that before."

After a brief pause, Dick said, "We had planned to ride about three hundred miles each day."

Rob broke in, "Yes, but we covered 424 miles that first day and stayed all night in the Kansas City area at a roach-infested motel. The room looked clean but when I turned the light on about midnight, the roaches scattered in every direction. There must had been thousands of them. They gave me the jitters! I hardy slept the rest of the night, thinking about them getting in bed with me. You didn't know it, but I left the light on the rest of the night."

Dick replied, "You should had been as tired as I was, I didn't know anything about the roaches until you told me the next morning."

"Yeah, you slept like a hibernating ground hog." Rob related. "The next morning we ate a late breakfast near Topeka and the temperature soon climbed into the nineties. By mid-afternoon the pavement was so hot that it looked like we were riding into a shimmering lake of water."

"Right," Dick chimed in, "And the parched ground, was barren for miles. Occasionally, we would see an abandoned house or barn with soil drifted up to their eves. That area was part of the great dust bowl a few years before."

Excitedly, Rod exclaimed, "That was when it all happened. Our cycles went out of control. It was like we were on black ice. Applying the brakes only made it worse. Both of our cycles went down at the same time, sliding across the highway. We ended up astraddle of our bikes on the shoulder on the road."

Dick nudged in, "I remember seeing a sign a

mile or so back; 'GRASSHOPPER AREA', but never dreamed that there would be anything like we encountered that day. There we were, dripping wet in bug goop, thirty miles from nowhere. The highway was a slimy mess the next twenty miles." Fortunately, neither of us were injured."

"Insects! Millions of them! Immediately we were covered with throngs of leaping, jumping grasshoppers." Ron related. "Clouds of them. It was like twilight. We could barely see each other. To talk we had to yell to be heard above their loud droning. We fought them off and they just kept coming in swarms.

"What I'm going to tell you is unbelievable, but it's the God's truth. Snowplows had been used to clear the roadway and dead grasshoppers were piled up to a depth of three feet along the sides. Road oil couldn't have been any slicker. After several hours of sliding back and forth trying to keep our cycles upright we were completely exhausted. Our bikes were a gummy mess. Hopper juice was in every crevice and the chain guard on both bikes was jammed with dead grasshoppers.

"Between the heat and the grasshoppers, we only traveled about 120 miles that second day."

Dick, clearing his throat, finished their story. "After a day of cleaning up our bikes, we rode north into Nebraska and the Dakotas where we enjoyed cooler weather, returning home by the weekend. We never made it to Pike's Peak as the grasshoppers had done in our enthusiasm. Maybe, after the war, we'll take our wives and tour the western states."

LST-1107 Earned no Battle Star

LST-1107 the 164th keel, laid down on 13 December 1944 at Evansville, Ind., by the Missouri Valley Bridge and Iron Co.; launched on 29 January 1945; sponsored by Mrs. Frank Parks; and commissioned on 21 February 1945, Lt. G.P. Lynch, in command.

Following World War II, she performed occupation duty in the Far East until mid-October 1945.

She was decommissioned on 1 May 1946 and struck from the Navy list on 3 July that same year.

On 28 February 1947, the ship was sold to the Higgins, Inc., for their operation.

LST-1107 did not earned a battle star for World War II service.

Chapter

41

Depression Survivors

SHIPYARD WORKERS WERE THE GENERATION WHO SURVIVED THE GREAT DEPRESSION WHEN COMMON SENSE WAS STILL BEING USED IN AMERICA.

Dike Jones was one of those workers that came from every corner of the nation for employment in the Evansville area. There was no doubt in his mind that we could win the war. His forebears were among the emigrants that opened the western frontier and traveled the Oregon Trail as they left St. Louis, in the summer of 1846, to join fellow travelers in Leavenworth, Kansas.

Klondike was Dike's real name. He said that he had been told he resembled his grandfather, a tall, strong figure, with clear blue eyes, and broad cheekbones hidden under red stubble sideburns.

His parents had related stories to him of how they forged rivers, and made friends with the Indians, and how his grandfather had even smoked the peace pipe several times with Indian chiefs along the trail.

Dike told how he and his wife, Dolly, and their five children had endured the Kansas dust storms during the 1930s' and the harsh frigid winters that swept the Midwest.

He was the typical type of worker who made up the civilian work force in the defense plants all across America. They were the patriotic workers who could be depended upon to give their undivided attention to the job they were asked to do.

Ethel Hicks was a welder at the shipyards, whose true blue, dyed in the wool ancestors, had followed Daniel Boone, as he blazed the trail through the Cumberland Gap into Kentucky. Ethel and Dike, along with thousands of depression survivors, made up the invasion force on the home front.

Thousands of patriotic Americans signed up to purchase War Bonds through a payroll savings plan. An eighteen dollar and seventy-five cents bond yielded twenty-five dollars at the end of ten years. It was a great way to save for the future and it also helped the war effort as the government attempted to spread the cost of the war over a period of ten years.

Price controls had been established by the Roosevelt Administration even before the war was declared. Later when meat was rationed as well as sugar, along with so many other things, made them realize that it was going to be a tough fight. Many people planted Victory Gardens in their front yard, with beans and tomatoes growing among the roses.

Everyone accepted the rationing of gasoline when it went into effect on May 15, 1942. They willingly honored the national speed limit of thirty-five miles per hour in order to save rubber tires.

Hardship often brings out the best in us, and so it was with the depression survivors.

Chapter

42

The Big Flood of 1945

THE RAINS CAME IN LATE FEBRUARY AND CONTINUED FOR DAYS UNTIL NEARLY EVERY HIGHWAY IN THE TRI-STATE AREA WAS CLOSED BECAUSE OF HIGH WATER.

Several streets in the city of Evansville were flooded. The backwater of the Ohio River raised Pigeon Creek out of its banks, flooding all of the low-lying areas. The stench was nauseating near the West Franklin Street Bridge where hundreds of dead fish and bloated river rats were visible floating in the polluted backwater slough of the creek. Manhole covers, on East Columbia Street, rattled, vibrated, and popped, producing a babbling sound as storm water and raw sewerage were released overflowing into the street.

According to the "*Evansville Courier*" on Wednesday, March 7, 1945 the shipyards were closed and would be until Monday, March 12. As I remember, the shipyards closed at the end of the day shift on Tuesday, March 6, as Highway 41 from Henderson, Kentucky was flooded and many workers were unable to report for work.

Although shipyard production was shut down, small LCVP lifeboats off the LSTs anchored at the docks were provided with crews to help rescue people from their flooded homes.

Servel Corporation, where my wife, Edna worked, was also closed, so we decided to drive up to my parents' home in Washington County, Indiana, to get out of the flooded area. That turned out to be a foolish idea. The flood was everywhere.

We remembered that the French Lick, West Baden area always got flooded during extreme wet seasons, so we decided to drive up State Road 57 to Washington, Indiana, and then take US 50 east.

Twenty miles north on Highway 57 we ran into a stretch of standing water about six inches deep that extended for several hundred feet. It was backwater from the upper end of Pigeon Creek, just south of the community of Buckskin.

About four miles north of Oakland City, the Patoka River and neighboring tributaries were on a rampage. Once we splashed through that, the rest of the roadway was fairly clear.

Some five miles north of Petersburg, we crossed the White River Bridge where the river bottoms looked like one gigantic muddy lake.

At Washington, Indiana, we turned right onto US 50, following the natural scenic highway east through Montgomery, Loogootee, past Jug Rock at Shoals, to the junction of State Highway 60 where we left US 50 and continued east towards Mitchell. All along the highway, fields were flooded, with thousands and thousands of acres inundated.

East of Mitchell on Highway 60, we incurred no major water problems until after we passed the old hairpin curve entrance to Spring Mill State Park. At that point, the old highway veered back west over the roller coaster terrain. After a shower, every low place became a big puddle in the middle of the highway. The intersection at Liberty Church and Orleans Road was flooded beyond forging.

Edna and I both became concerned when we had been on the road over four hours. It was getting dark early with storm clouds producing flashes of lightening that lit up the sky. We had driven through shallow pools of water in a dozen places and even had water splashing up into the interior floor-boards of our 1939 Ford sedan. We were at our wits end as to which road to take. Every road seemed to be blocked by high water as additional showers fell.

That was when the Lord answered our prayers. I know that I was praying for a solution, and I'm quite certain that my wife Edna, was saying a prayer. How were we going to get to my parents' home north of Campbellsburg?

Divine Guidance from the Lord, definitely took over. Divine Guidance was the forerunner of the present day GPS receiver.

At one place we turned the car around and backtracked a short distance to another road. By following that road, which was on higher ground, free of ponds, we arrived safely at my parents' home. It had taken five hours to travel the distance from Evansville, which was normally a three and a half-hour drive at 35 mph. *(WWII national speed limit)*

Years later, when I tried to retrace the route that we drove that stormy March night, I was never able to determine which road the took after leaving the Liberty Church crossroads in Orange County.

The following Sunday afternoon, with the floodwaters receding, we returned to our apartment on West Broadway in Evansville. I reported back to work at midnight on the Graveyard Shift to resume work on Monday morning, March 12, 1945.

Within the hour, shipyard sounds reverberated across the city. Once again, giant flood lights illuminated the sky as burners, chippers, maintenance, office personnel, outfitters, painters, riggers, ship fitters, shipwrights, tank testers, and welders resumed work with the determination and zeal that two years earlier had earned them the coveted Army-Navy 'E' Award for Excellence in War Production.

~~~~~~~TAKE A BREAK ~~~~~~~

Two children were watching a motorboat pull a man on skis down the Ohio River, "Why is that boat going so fast?" Asked five-year-old Lucy.

"It's because that man on the water behind the boat is chasing it...Dummy!" Her brother shouted!
~~~~~~~

Chapter

43

Spirit Making & The Law

EVANSVILLE SHIPYARDS WAS A LONG WAYS FROM THE MOUNTAINS OF EASTERN KENTUCKY, WHERE ONE OF THESE STORIES ACTUALLY HAPPENED ACCORDING TO MY INFORMANT.

Al Berry, a husky outdoorsman of many talents, was a native of Leslie County, Kentucky. He had a checkered career. He had been an informer, a still operator, a school bus driver, and a sheriff before becoming a welder at the shipyards.

He grew up in the shadow of the mountains seven miles south of Hyden, where his Pappy was a revenuer for years. Al quit school at age twelve and helped his father as an undercover agent. He bragged about having over twenty years experience in the still game. Al claimed that from several miles away he could smell a still that was cooking off a batch. Later in life, he claimed to have made many trips into Kings Port, Johnson City, and Knoxville, Tennessee, delivering the mountain spirits.

"As a kid," Al said, "I hunted squirrels all year. It was always good hunting in the mountains up above the Middle Fork of Kentucky River. I also spent lots of time roaming the woods and mountain trails around the Stinnett Creek area. If I found a still, I'd tell my Father, who was a Federal Agent.

"While packing a 12-gauge Winchester with a few squirrels hung on my belt, no one suspected that I was doing anything other than hunting. Occasionally, I would accidentally stumble onto a still hid deep in a mountain ravine next to a spring. I became a target of suspicion after half a dozen operations had been seized by the Feds and arrests made."

Al continued, "A few years later, my father and I got caught in a crossfire shoot out with moon-shiners. My father was killed. That was when I joined up with two other guys, Chuck and Hank, in order to save my own life. They operated a still up Stinnett Creek and I was their tripper making deliveries. I say; if you can't fight them, join them.

"After I married, I cut timber and drove a school bus. I don't think that anyone suspected that I was anything other than a law-abiding citizen.

"Later, I was elected High Sheriff and arrested a few of my former buddies, including Chuck and Hank. For that reason alone, I can never go back home."

A few nights later during one of our breaks from welding, Al related another story. "Back when I was a teenager, involved with Chuck and Hank, we operated a still up Stinnett Creek. One afternoon two strange men stumbled into camp carrying a couple of squirrels and asking directions how to get back to highway U.S. 421 and civilization. They had on old clothes and talked like old friends, but there was one thing that I noticed that made me very suspicious. I just knew they were revenuers.

"I figured that they would want to taste a sample of our makings, so I offered each a cup of corn-licker into which I dropped a Mickey. It didn't take long to take effect on them and they sat down on a log and soon were out like a light. We tied them up back to back, and removed their new Wolverine work shoes. Those shoes had given them away."

Al chuckled as he recalled the incident that had happened years before. "We started dismantling the still and every time they stirred, Chuck would force more liquor down them. Hank went over the mountain and came back with his uncle's two burros, which we loaded down with our equipment. By the way, those Wolverine shoes were a perfect fit for Chuck and me."

Al continued with a big smile as he recalled the situation, "As we were ready to leave, we loosened the ropes a little which we had used to tie them up with and forced more whiskey down them. That would keep them safe for the next twelve to fifteen hours. In other words they were dead drunk!

"Around two o'clock the following afternoon, two of the sorriest looking specimens to be humans came limping down the trail in their sock feet. They sat down on the bank of the Middle Fork Kentucky River soothing their aching feet in the cool waters. The river runs parallel with highway US 421, right near the general store, which was also the U.S. Post Office in Stinnett, Kentucky."

Al's story presented so many facts that it would be difficult for anyone to have just made it up. I believe he told it like it was.

Another fellow worker, Jim Burk had a different kind of moonshine story. From the way he told it, evidently it happened over in the hills of southern Indiana and not too far from Evansville.

"Back in the nineteen twenties and early nineteen thirties, few grocery stores had milk for sale. Dairy farmers would have a few dozen cows and bottled the cow's milk in quart bottles. Then they would deliver the milk to their customers early each morning as they made their rounds in an enclosed one-horse wagon."

Jim continued with his tale, "This particular enterprising young farmer occasionally operated a moon-shine still. It was well hidden down by the spring where he cooled his milk in ten gallon cans overnight.

"He painted half a dozen quart milk bottles white, which he filled with his home brew of 'white lightening.' He had a few special customers who would call him for whiskey. He would make their special delivery while making his daily milk rounds.

"Evidently, the fellow figured out how to cover up the odor of his whiskey making. Once a week he trucked distillery mash from Cook's Brewery at 8th and Main in Evansville to his farm, feeding the malt-grain millings to his fattening pigs."

For several years his method worked well until one foggy morning a few years ago when he became a little confused and delivered two painted quarts to the county prosecutor's home. That put a quietus to his operation plus serving time in the pen.

Chapter

44

Echo Off The Bulkheads

MANY STORIES ECHOED FROM THE BULKHEADS DURING MY THREE YEARS OF WELDING UP THE LANDING SHIP TANKS.

Connie and I work as a welding team on the skeg crew, welding drive shaft tubes into the cast iron rudders. Connie's, real name was Conrad. He was at least thirty some years older than I was at age 21. Connie mentioned the fact that he had served in the merchant marines for 20 years. He claimed that he had been around the world six times. From his conversation, he was one of those old diehard sailors with a lover in every port.

Connie talked about the many countries and ports that he had visited, including the enchanting names of the old world such as Bangkok, the capital of the kingdom of Siam. Other places that he talked about were Shanghai, Manila, Borneo, Sidney, Timbuktu, Tripoli, Morocco and the Barbary Coast.

His claim that he had met Blackbeard the Pirate who lived back in the seventeenth century and his story about working on an assembly line building Studebaker cars in Detroit had me doubting his truthfulness. The Studebaker Company had only one factory and it was in South Bend, Indiana, not in Detroit, Michigan.

We were part of the graveyard shift skeg crew, welding the stainless steel drive shaft tubes into the cast iron rudder mounts on the aft end of the LST's. Each ship had two drive shaft tubes.

It was an interesting job, however, a very difficult welding job. There was no way that two welders could complete the welding on one drive shaft tube in an eight-hour shift.

On this particle night the temperature had dropped to ten degrees Fahrenheit. Two salamander kerosene heaters; one placed on each side of the cast iron skeg rudder was necessary to keep the metal warm enough so that the welds would not crack.

We stood on a scaffold some eight to nine feet above the cement in order to be at a work level of the tube. The welders worked in teams welding on opposite sides of the twelve-inch diameter stainless steel drive shaft tube.

After each weld pass, a chipper was required to peen the weld with an air hammer to relieve the stress in the metal where a deep groove required many weld passes in order to be filled.

At times, we would have to wait for half an hour or more for the roving chipper to show up to peen the weld clean. That was when Connie unloaded his marital problems on me.

"Bea and I met at a bar one night three years ago out north of town on Highway 41. We got married the following week. She told me she had one husband deceased and had been divorced twice."

I could barely hear what he was saying above the noise of the air hammer as it peened the weld.

After the noise subsided, Connie cleared his throat and continued talking about his problems.

"Things just ain't like it used to be with Bea and me. I began to suspect there was more about Bea's past life than I knew about. An old friend told me that I had better get a detective and have him check on Bea. So I did, and yesterday morning the detective told me that according to the death and divorce records in the Marion County courthouse in Indianapolis, Bea had been divorced once, but had buried five husbands, and I don't intend to be number six!"

The following night the temperature dropped to twelve degrees below zero. Our foreman told us to not do any welding, as the welds would cool so fast that they would crack. In fact it was so cold that we didn't do much talking. A tarpon was nailed to the wood scaffolding for a windbreak. Where it was fastened to the guardrail it bagged out like a hammock. We sat on the two by four railing with our backs against the canvas and dozed off to sleep.

That was the last time I saw Connie. Evidently, he quit his job and took off for parts unknown. Maybe he left to follow his adventurous dream of going to Paris and joining the Foreign Legion although he was getting a little old for that.

On another occasion Connie had commented, "Before I kick the bucket, I want to ride on the 'Orient Express'."

As this story echoes through my mind, I wonder if Connie is still out there somewhere riding the 'Orient Express'.

LST-1108 Earned no Battle Star

LST-1108 the 165th keel laid down on 16 December 1944 at Evansville, Ind., by the Missouri Valley Bridge and Iron Co.; launched on 1 February 1945; sponsored by Mrs. Edward H. Barnard; and commissioned on 27 February 1945, Lt. C. V. Lieb, in command.

Following World War II, LST-1108 performed occupation duty in the Far East until early December 1945.

She was decommissioned on 15 August 1946 and struck from the Navy list on 25 September that same year.

On 10 January 1948, the ship was sold to the Argentine Navy where they renamed LST-1108 as Cabo San Vicente BDT-14

LST-1108 did not earned a battle star for World War II service.

Chapter

45

Roving Traffic Inspector

HAVING OVERSLEPT, I HAD TWENTY MINUTES TO GET TO THE SHIPYARDS AND CLOCK IN BEFORE THE SHIFT CHANGED.

Our apartment out on East Broadway was about a mile and a half from my usual shipyards parking spot and then I would have a quarter mile to run in order to get to Clock Alley # 2.

I gave my recent bride, Edna, a warm hug and a lingering kiss before pulling on my work clothes. I grabbed my car keys and ran down the back stairs, jumping into the 1939 Ford 2-door sedan that I had purchased about a month earlier.

As I approached the intersection at Broadway and Barker Avenue at thirty miles per hour, I turned on the left turn signal and started to turn as I had the right of away. A car facing me at the intersection had a stop sign, but failed to recognize it. With my quick reaction, we missed colliding by mere inches.

As I looked into my rearview mirror, I saw that the car had turned and was overtaking me at a high rate of speed. I figured that it was a drunk or a mad teenager, so I rolled up to fifty and took a quick right turn going up the hill pass Reitz High School. When I didn't see headlights behind me, I figured I had lost him. Once over the hill, I meandered down several short side streets towards the Meade Johnson buildings to where I usually parked.

As I jumped out, a 1941 '98' Oldsmobile Hard Top 2-dr. pulled up along side with the passenger window down. The driver was yelling at me. He was an older man in a business suit.

"You nearly caused me to have a wreck and I have been looking for that car ever since it was involved in a hit and run accident over two months ago. Also, it fits the description of a car that was used in a bank robbery last year. I'll have it pulled in and you will have to come down to City Hall in the morning as I'm filing charges against you!"

As if I didn't have enough problems trying to get to work, then this crazy drunk had to show up. What was he talking about, a hit and run accident?

I had recently bought the car as is, with the right rear fender smashed, from an Used Car Lot on North Main Street. Who was this old guy anyway?

I ran on towards the Clock Alley with more worries than I needed at midnight. As soon as I clocked in, I stepped over to the guard office and told the officer on duty about my recent experience and described the vehicle and driver. He replied; "That's the city's traffic inspector. We've had problems with him before. Don't worry. Your car will be safe. I'll have a guard keep check on it."

The next morning, as I got off from work, my car was parked where I had left it. I never heard any more about the hit and run charges or about the car having been involved in a bank robbery.

Chapter

46

The Stone Man Mystery

EVANSVILLE MADE NATIONAL NEWS ON OCTOBER 1, 1902, WHEN A LOCAL NEWS PAPER, *THE DAILY JOURNAL*, PUBLISHED, *"PETRIFIED BODY OF MAN IS FOUND"*

"By 1902, the river had 'receded down to its rock, sand, and gravel bottom'...that fall.... dredging the Evansville waterfront.... the workers, according to local press accounts, discovered the body of a small man, seemly turned to stone."

"A reporter for the Evansville Daily Journal arrived in time to view the object... He filed a news story describing the figure as 'five and a half feet in height with a powerful chest' and with clearly distinguishable facial features."

Description of the stone man as given in the newspaper, could very well have been the author's great, great uncle, Jacob Lee, a steamboat pilot, who disappeared in the year of 1857 at the age of 45, and was never heard from after that date.

In a conversation with my grandmother, Louisa Jane Lee Cornwell, years ago when I was a teenager, she had described her uncle, Jacob, as being five foot and five inches in height and weighting one hundred forty pounds. She said, he walked with a limp, after having infantile paralysis,

leaving one leg shorter than the other. Today the disease is referred to as polio.

"On October 3, Indiana Governor Winfield Durbin, accompanied by his staff and several friends, journeyed south from Indianapolis to view the astonishing specimen. The governor was not content merely to view the body; the press reported he critically examined every part of it. Afterward, Governor Durbin announced: 'there is no doubt but that that specimen is the genuine petrified body of a man. It is the most wonderful thing I have ever seen, and I am greatly interested in it.'"

The Evansville Courier reported, *"the theory is advanced that the image is probably the remains of some relative of the inquirer who had mysteriously disappeared."* **They also published, a long story about an impoverished widow who thought the recently discovered petrified man was her missing spouse.**

'That is my husband. [T]here is not the slightest doubt about it.' "The woman who claimed the stone image as her "lost" husband ended several years of living with anger and shame at the possibility that he had simply left her."

A number of different stories surfaced the following month, each recalling missing men who had disappeared or had drown many years before,

To add a little humor to the story, *"the Courier, a Democratic newspaper, also printed the observation that the 'petrified man' was in reality the body of 'a petrified republican office holder' found in the basement of the courthouse."*

The *Evansville Daily Journal* reported on October 17, *"the three hundred pound stone man is missing."*

Two weeks later, November 3, the *Evansville Courier* ran a story. *"It is reported, the stone man is on display at Rockport. Soon after that, the stone man seemed to have vanished, as there was no other record of its whereabouts."*

The answer to the dark mystery of the stone man may very well lie buried under the 12 acres of concrete that once was the LST shipyards.

In 1942, as the forty-one acres along the Ohio River were being graded in preparation to construct the Navy Ship Yards, bulldozers buried large concrete chunks and other debris in the area. One heavy object resembled a discarded cemetery statute. Could that have possibly been the stone man? As it had been forty years since the Evansville Petrified Man discovery, most everyone was too young to have remembered anything about the story. Only the famous, *Angela Lansbury*, as amateur detective, *Jessica Fletcher*, in the TV series, "*Murder She Wrote*" could have possibly solved the mystery. DNA testing was not available back at that time so the stone man's identity may forever remain unknown.

Note: *The above quotes, are brief excerpts from the story entitled; "The Most Wonderful Thing I Have Ever Seen: Indiana's Contribution to Petrified Man Hoaxes", by Carl Runyon and Randy Mills, as it appeared in the Indiana Magazine Of History, Volume 104, Number 4, December 2008.* Reprinted by permission.

LST-1109 Earned no Battle Star

LST-1109 the 166th keel, laid down on 21 December 1944 at Evansville, Ind., by the Missouri Valley Bridge and Iron Co.; launched on 6 February 1945; sponsored by Mrs. Paul H. Derrick; and commissioned on 28 February 1945, Lt. William A. McCaskill, USNR, in command.

Following World War II, LST-1109 performed occupation duty in the Far East until early November 1945.

She was decommissioned on 6 May 1946 and struck from the Navy list on 19 June that same year.

On 13 June 1948, the ship was sold to the Walter W. Johnson Co., for scrapping.

LST-1109 did not earned a battle star for World War II service.

Chapter

47

Blowing Straw Stacks

ALONG ABOUT THE SECOND WEEK OF JULY IN 1945, THE EVANSVILLE NAVY SHIPYARDS WERE DOWN TO MAKING BARGES AND FINISHING UP FOUR LUXURY YF BOATS (Freight and Ammunition Lighter).

All of the LSTs, (Landing Ship Tanks) had been launched with only a few still at the outfitting docks.

Over three-quarters of the employees had checked out for other jobs. I turned in my badge, clock # 4214, hoping for a change of pace.

The European War was over. It would only be a matter of time before we 'slapped the Japs.'

Edna and I packed up all of our personal belongings in our 1939 Ford 2-dr. sedan. Saying good-by to our friends, Eddie and Genevieve Lynn from whom we had rented an upstairs apartment, we left Evansville. We had purchased an eighty-one acre farm north of Campbellsburg, Indiana, near my parents. We would need to make arrangements for someone with a truck to move our furniture.

On our way, we stopped at Edna's sister's home, Ruby and Orval Katterhenry, who lived west of Huntingburg, Indiana, near the Dubois and Pike County line.

Orval was starting the threshing season and needed some help, so we decided to stay a few days. Getting up at 4:00 A.M. was fun, but I ran out of energy long before sun down. Then usually, we had to move to the next job site before calling it a day. About ten o'clock P.M., Orval would head the old International truck homeward.

After a bite to eat and a quick wash-down, the straw tick felt mighty good for a few minutes. The next thing I knew, the alarm clock was going off. I struggled into a clean long sleeved shirt, downed some coffee, eggs and bacon for breakfast and met Orval out at the truck ready for another hot day with the threshing gang. July weather is usually hot and dry by the third week and this was no exception.

Some days we thrashed at two different farms in the same community. One afternoon we did a job for the Hunfeld Brothers, Willie and Walter. They were up in years and neither had ever married. They spoke broken English mixed with a German accent. I had difficulty understanding them.

My job was to operate the thresher blower cap for the straw stack and oil the Babbitt bearings every fifteen minutes while blowing hayseeds and dust out of my nose and eyes.

Most of the farmers had winter wheat to be thrashed. A few had spring oats. The Hunfeld Brothers had a sizeable acreage of bearded wheat to be thrashed.

Bearded wheat sheds off a lot of sharp sticky thistles like needles. You can imagine how miserable Orval and I felt. My shirt was drenched with perspiration and pricked with dozens of thistles down my back, neck, and even into my shoes. My every move was painful.

However, having been an electric arc welder at the shipyards for three years, I was used to working in undesirable conditions. I still carry welding burn scars after all of these years.

Standing on the top rear of the threshing machine operating the blower cap, I cross-stitched the straw. I flipped the blower cap up and down from left to right for one layer of straw and then the next layer was laid crossways from the farthest side to the side next to the threshing machine.

So the straw stack grew as the afternoon progressed.

About every foot in height, I would turn the blower cap upside down and let the straw mist a layer over the entire stack.

Orval had instructed me to mist the top of the straw stack with the final load of wheat bundles being thrashed to seal the top of the stack so rain would shed off. To do otherwise was my idea.

Having blown several stacks in the past few days, I felt that I had mastered the job. It was time to relax and experiment a little on this stack. I kept the perimeter, or outer boundary, of this stack higher as it progressed skyward.

I figured that this would have an interlocking effect, helping to keep the stack from sliding off sideways later. All the while I continued to weave each layer back and forth, and then cross weave the next layer, side to side. Every foot in height I turned the blower cap upside down, misting a heavy layer of straw over the entire stack.

About the middle of the afternoon, Orval yelled above the roar of the tractor and the thrasher . . . "SHUT HER DOWN BOYS! . . . WE'VE GOT A HOT BOX!"

Evidently, I had become so interested in blowing the straw stack that I failed to remember to oil the Babbitt bearings.

While the other fellows headed for the shade, Orval and I had to pour a new Babbitt metal bearing and let it cool before we could thrash the rest of the crop. By the time Orval checked every thing out and I had oiled all of the other bearings, we were ready to start operations again. All ten-wagon teams in the field were loaded and lined up waiting on us.

The next year at the Satkamp Family reunion, Orval said that the German brothers had asked him to get that lad that blew their straw stack last year. "In all of their years of farming, that was the first stack that had not caved in as the cattle ate holes through it."

Chapter

48

Washday 'Blues' Without Electricity

MANY RURAL FAMILIES IN THE UNITED STATES DID NOT HAVE ELECTRIC SERVICE UNTIL AFTER THE WAR.

My parents were one of those unfortunate families. Rural Electric Membership Corporation was available in southern Indiana within two miles of their home, however, when war was declared on December 8, 1941, line construction was cancelled.

Saturday was Mother's washday. On a weekend, when the shipyards didn't work, I would go home to help her do the wash. Washing clothes was a form of religion with my mother, bless her soul.

Mother didn't have much faith in washing powders like Borax, so she would shave a bar of ivory soap into a big copper wash boiler filled with cistern water from the kitchen pitcher pump.

All of the white clothes, including sheets, pillowcases, and white shirts had to be immersed and steeped in boiling water. The water was heated on the old kitchen range. Mother usually set the alarm clock for one hour to be sure that they stayed in the wash boiler the full length of time.

On cold winter days that wasn't so bad. However, in the summer, when the temperature was ninety degrees in the shade, the heat from the wood and coal-burning cook stove ran the thermometer above the triple digits and melted the lard in the pantry.

Once the steaming mass was fished out with a bleached mop handle onto a boiler lid, it was rushed out to the Maytag washer, located on the enclosed back porch, to be rinsed for the first time.

For those of you with no knowledge of the operation of a gasoline washing machine before the day of electricity, I'll try to explain.

First, purchase two gallons of *'High Test Ethel'* gasoline. Fill the special Maytag one-quart can, and add about two tablespoons of Maytag motor oil. Then pour the mixture into the gas tank in the base of the motor.

It is best that the spark plug be removed and cleaned before attempting to start the motor. At times, the motor will refuse to start and act as if it is in cahoots with the devil. When that happens, sometimes it helps to remove the spark plug and hold a lighted kitchen match to the base, heating it up, and then quickly replace it in the motor block.

The cast iron starter lever has to be stepped on with all of one's weight in order to turn the flywheel. No battery is needed as the spark plug located above the one cylinder piston receives its spark from the magneto in the flywheel.

A mini carburetor provides a vapor of gasoline that explodes with a loud bang of blue smoke through the flexible exhaust pipe stuck out an open door. Now the Maytag is running!

It took someone with a lot of stamina to kick-start that Maytag two-cycle gasoline motor, let alone to keep it running.

Once the Maytag decided to run, we knew the wash was underway and all would be well until one of us ran out of gas, and usually it was the motor.

I can still hear that old Maytag washing machine banging away as it did the laundry.

The second rinse was into a large galvanized tub filled with clean cold water. After going through the wringer the second time, the wash was ready to be hung on a solar clothesline outdoors.

The clothesline consisted of a number nine wire stretched from the corner of the chicken house to the nearest corner of the barn, a distance of sixty some feet. Near the middle of the line was a forked prop pole. It was needed to raise and tighten the clothesline to make it more stable.

No matter what the weather was like…come rain, sleet, or snow, the laundry had to be done. I certainly didn't look forward to the Saturday washday ritual. One Saturday the Maytag refused to run and we were all day doing the laundry on the old hand rub washboard, one piece at a time. That was when we found ourselves with shriveled fingers, hanging up clothes by moonlight.

There was no such thing as permanent press clothing during those war years, and if you wanted to look presentable, shirt collars and cuffs had to be starched and then ironed smooth without a wrinkle.

Diapers needed to be ironed. It was common knowledge that babies diapered with an un-ironed piece of cloth were usually very fussy. A screaming baby might not be crying from having been stuck with a safety pin, but from a rough uncomfortable diaper.

I will assure you that standing over an ironing board all day long wasn't any fun. A lot of sweat and toil from the labors of keeping a family in clean and neatly ironed clothes caused many an aching back as well as a splitting headache.

Those without electricity did their ironing the old fashion way by heating irons on the wood and coal cook stove with a hefty fire in the firebox.

My mother thought, "If I only had a gasoline iron, it would make ironing so much easier."

Salesmen glorified the process of ironing with a gas iron but gave no warning of the danger. Most housewives found the operating instructions rather complicated. The man of the house had the responsibility of keeping the gasoline iron operating. The teacup size gas tank was located next to the handle. The tank had to be kept under pressure with a small hand pump. To heat the iron, a match was struck to ignite the burner in the base, which when lit made a loud sizzling sound. A knob at the rear of the iron controlled the temperature. Once the iron was hot, it was ready to use. Occasionally, the iron would flare up, which was a dangerous factor. From my personal experience, they were not safe to use.

Living without electricity meant that you had to do every thing the hard old fashion way.

Chapter

49

Horseplay Ends In Death

The passing of time has a way of obliterating the memory and the evidence of what possibly was a crime committed three score and seven years ago.

In the fall of 2009, while on a guided tour of the LST 325 Memorial at Evansville, my memory was refreshed in more ways than I had expected. I was soon reminded of the need to wear my hard hat after bumping my head as I entered the galley. It is necessary to duck and step high over the threshold while passing through all watertight bulkhead doors.

About midway through the tour, a gray headed man about my age introduced himself as Al Oar. He said that he had worked at the shipyards during the summer and early fall of 1942. In discussing our experiences, Al told me a story relating the tragic and mysterious death of one shipyard worker who had fallen off the main deck of an LST while it was under construction on the Ways.

Mr. Oar, who is now eighty-five, but was only eighteen at the time of the tragedy, described the scene. "I well remember. It was the last Friday in September 1942. The morning arrived with a cool breeze, overcast sky, and signs of a possible rain later in the day. The constructions of the Landing Ship Tanks were progressing smoothly."

Al continued, "These two guys had been horsing around. I was standing next to this co-worker and I saw what he did to the other man and heard him say in a low voice to himself, 'Guess I shouldn't have done that.'"

"The horrifying screams from the victim I can still hear, as they were drowned out amidst the tremendous ear-piercing clash from several metal chipping air hammers, and the throbbing vibration of the gantry-crane. The man fell some thirty-five feet to his death on the concrete below. That really frightened me." Al related. "It was a foolish prank that his supposedly friend did to him. I was thankful that I wasn't personally involved. I was only eighteen years old and was horrified by what I saw happen. That was the last day I worked at the shipyards, as I had to report for military induction the following Monday. I didn't tell anyone about it. Even to this day, its memory lingers with me."

It was a few days later, before I heard about the man falling to his death. I was working on another ship nearby at the time. No one seemly knew what caused the worker to fall. The coroners report coldly stated that the man died from a fall at the shipyards. His death was classified as a war-casualty on the home front. There was no apparent reason for the man to have taken his own life. He was in his early twenties.

However, what Al's memory now reveals cannot be awarded a court hearing.

If an attorney had handled this case, he would raise a host of legal questions. People's perceptions and memories sometimes get confused. Sit in any courtroom and listen to an attorney cross-examine a witness. Eyewitness testimony isn't always to be trusted.

According to Al, there were only three people who actually knew what had occurred, and each kept silent.

The victim couldn't tell his side of the story.

The worker who was responsible for the man's death wasn't talking, and Al who now claims to have been an eyewitness was so frightened at the time, that he didn't tell anyone. From the information that Al presented, the victim didn't commit suicide.

If the victim had been provoked to the point that he accidentally fell or jumped overboard, then Al's co-worker could have been charged with murder or at the very least, manslaughter.

Might Al, as a witness to the fact, be disqualified on the grounds of his sustained imagination or better still be accused of his total distortion of the actual fact when he claims the provoker goosed his fellow worker causing him to jump, which resulted in his death?

One thing is for certain. The man that died never expected to die that day when he woke up that morning...to not see his tomorrow!

LST-1110, Earned One Battle Star

LST-1110, the 167th., keel was laid down on 28 December 1944 at Evansville, Ind., by the Missouri Valley Bridge & Iron Co.; launched on 9 February 1945; sponsored by Mrs. Sydney Kolb; and commissioned on 7 March 1945; Lt. Alton S. Lee, USNR, in command.

During World War II, LST-1110 was assigned to the Asiatic-Pacific theater and participated in the assault and occupation of Okinawa Gunto in June 1945. Following the war, she performed occupation duty in the Far East and saw service in China until late January 1947.

During the succeeding 10 years, LST-1110 engaged in artic re-supply operations, operating out of ports on the California coast.

On 1 July 1955 she redesignated as San Bernardino County (LST-1110) (q.v.) after a county in California. She was decommissioned on 15 August 1958.

San Bernardino County was struck from the Navy list on 6 February 1959 and was transferred to the Republic of China Navy. They renamed it as Chung Chiang (LST-225).

LST-1110 earned one battle star for World War II service.

Chapter

50

Final Thoughts...

EVERY GOOD THING HAS TO COME TO AN END, AS CAROL BURNETT SANG AT THE CLOSE OF HER TELEVISED PROGRAMS.

I haven't run out of shipyard musings to relate, but it's time to close the curtain on this publication. These stories represent a simpler time in America's history, even during a war, when solid family value and homespun virtues prevailed.

It has been a rewarding trip down memory lane on each of the foregoing episodes. During the writing of these stories I have learned that a person needs long-range goals to keep from being discouraged and frustrated. Clear your mind of the term "I can't" and remember that through Christ, all things are possible.

My employment by the Missouri Valley Bridge & Iron Co., at the U. S. Navy Shipyards in Evansville, Indiana, during World War II, was a formal education. The military had rejected me, as I was thirty pounds underweight caused by an open skull fracture, which occurred at the age of 15. I am thankful that I was able to aid the war effort while making something worthwhile out of my life.

As you reflect back on these stories, please understand that history must be judged in the context of the era and location where the events took place. Also, customs and behavior can vary; however, the values of right and wrong will never change. Like Mark Twain said about his writings, "Portions of every story actually took place in one way or another!"

Weather was an important factor during the construction of the LST ships. Several stories relate to the weather conditions, as all construction was outside. The sizzling heat of the sun, building up on the steel, along with the heat from the welding was a challenge to all of the workers. The other challenge was to not freeze when the thermometer dropped near or below zero during the winter.

Life in the shipyards was filled with variety. Included are two short stories pertaining to love and romance and also two stories about fishing. The missing Petrified Man for those with a flare for mystery. Then there's the honest bank robber confession, along with a historic record of the down to earth workers who cussed, sweated and labored in constructing the ships into becoming the workhorse of the United States Navy.

During the period of my employment, my one suggestion to improve production of the LST boats was not recognized for a citation award, and I can well understand the reason why it wasn't. Immediately after I addressed the problem to my foreman, I was interrogated and suspected as being a spy and was under surveillance for more than a year. *(Chapter # 5 "Encounter With The FBI")*

The landscape of the Evansville riverfront has drastically been altered over the years, but there remains one lone monument dedicated to the memory of the World's Champion LST Builders. It is located on the 41-acre site along the banks of the beautiful Ohio River where at one time the shipyards rose up out of the bogs. The monument honors the toil, sweat, patriotism, and dedication of the men and women who helped build the LSTs, the work force of the United States Navy.

LST-325 was built in Philadelphia, PA.; commissioned 1 February 1943. However, in the 1944 Evansville Shipyard Pictorial Booklet, LST-325, in a Navy photo, is shown "land based" on the beach in Normandy, June 1944. The USS LST-325 Ship Memorial Inc., is now located at 840 LST Drive, Evansville, Indiana. How ironic. LST-325 is listed on the National Register of Historic Places.

LST-325 is being restored and is currently serving as a museum. Its mission is to educate the present and future generations about the role that Landing Ship Tanks played during WWII to keep America free and honor the men who bravely served their country aboard LSTs.

Evansville, Indiana, provided two of the greatest contributions to the nation's WWII effort in the battle for freedom. One was the LST amphibious boat and the other was the P-47 Thunderbolt plane.

During WWII, all of Indiana was located in the Central Time Zone with year round daylight saving time. There was no confusion with the need to change ten thousand clocks.

Let's keep the desire to fight for freedom alive in America today!

The last four LSTs type ships that were built at the Evansville shipyard were not listed as LSTs in the History of Landing Ship Tanks, Built at Evansville.

The shipyard workers knew no different, but that they were building another Landing Ship Tank, as the size and general appearance were the same as the other LSTs that had been built.

Evidently, it was some sort of a government secret why LST-1111, was redesignated AKS-16 and named Blackford (q.v.)

LST-1112, was redesignated AKS-17 and named Dorchester (q.v.)

LST-1113, was redesignated AKS-18 and named Kingman (q.v.)

LST-1114, was redesignated AKS-19 and named Presque Isle (q.v.) Later her name was changed to Vanderburgh (q.v.) Still later she was redesignated APB-45

According to New Word Dictionary of American English q.v. *[L. quod vide]* **The word** *quod* **is in reference to a prison or jail. The term** *vide***, is vision.**

So the LSTs redesignated as AKS (q.v.) evidently were built to transfer war prisoners.

One Nostalgia of the Shipyard can be found on the west riverbank for unloading barges.

GLOSSARY

The following Glossary lists words or phrases that were common expressions during the construction of the LSTs. However, they may be unfamiliar to today's readers. Many of the terms may have several different meanings. They are defined here only as they are used in the stories in this book.

Acronyms are made up of the initial letters of an official title. Abbreviations are used for names of organizations and government agencies. President Roosevelt's New Deal created some. There are dozens of them. LST for example, stood for Landing Ship Tanks. CCC, Civilian Conservation Core; NYA, National Youth Administration; NRA, National Recovery Act; WPA, Works Progress Administration; FBI, Federal Bureau of Investigation.

AKA: LST redesigned as Attack Cargo Ship.

AKE: LST redesigned as Advanced Dry Cargo Ship.

AKS: LST redesigned as Amphibious Cargo Ship or Supply Ship. General Store Ship.

APB: LST redesigned as Amphibious Self-Propelled Barracks Ship. Some were used as hospitals; remained parked at pier. Others were for transfer of war prisoners.

Anchor: A heavy steel or iron instrument with pointed arms as an upside down 'T' that is lowered from a ship to grasp the ocean, lake or river bottom to secure the ship from movement while stationary.

B -Tanks: They were ballast compartments. Many were used for fresh water holding and also diesel fuel. While navigating through heavy seas, other B-tanks were filled with seawater to increase weight.

Boat: Most any scalloped out form that floats, and can be used for transportation; from a canoe to an ocean going vessel.

Bow: The front part of a boat

Bulkhead: Any of the upright partitions separating parts of a ship.

C-Section: Compartments located immediately above the B-tanks on a LST boat.

Davits: A pair of upright arms that can be swung out over the side of a ship to lower a small craft.

Deck: Any horizontal structure in a ship serving as a floor such as a main deck, tank deck, top deck, poop deck, etc.

Dock: A landing pier or wharf. Usually a wooden platform extended from shore out over deep water to assist in transferring produce or material from the beach to the portside of a boat.

Dogs: Scrap iron brackets cut in the form of the letter "U" and tack welded to the deck plates so the shipfitters can attach a steamboat jack to pull sections of decks or compartments in alignment.

Gallery: The kitchen of a ship where food is served cafeteria style and often referred to as the mess hall.

Gun Turret: Is a stainless steel protective barrier around an anti-aircraft gun location on a warship.

Hatch: Any opening between decks of a ship that can be made watertight.

Jig: A pattern form on which something is to be assembled, maintaining the proper position in relationship to the finished product.

Keel: The center beam extending the entire length of a boat. The keel is the first section to be laid in the construction of a ship, which is considered the backbone of the boat.

Portside: Is the left side of a boat.
Starboard: Is the right side of a boat.

Quarters: Small compartments where the combat men sleep in three-tier hammock type beds. Ship crew quarters and officer's quarters usually have light springs and thin mattresses to sleep on.

Skeg, Skegg: The after part of the keel, or an extension of the keel upon which the rudderpost in mounted. **Skeg Welder:** A person that welds on a boat's stern.

Skids: They are subassembly locations on which several parts of the boat are assembled, forming a compartment or section, to be later attached to the rest of the ship.

Squelch Bulkheads: They are metal partitions in the stern of the ship with 18-inch diameter holes, offset every two feet, to stabilize the flow of liquids stored in that section.

Steamboat jack: It is a ratchet come-a-long jack, to pull metal deck sections and compartments together in alignment with each other.

Stern: The rear portion of a boat.

Tank: A compartment to hold liquid.

Tack: Spot Weld to hold the aligned sections in place until a finished weld can be made later.

Way: Location on which the ship keel is laid where the boat is constructed.

Wedge: A piece of wood or metal, thick at one end and narrowing down to a thin edge at the other, used for forcing something open or to tighten it up. A wedge can be used to force metal closer together or to drive a wedge between parts to separate.

Winch: A power driven machine, or hand operated with a crank, attached to a spindle on which a rope or cable is attached for hoisting, or lowering an item from one level to another level.

~~~~~~NOTES~~~~~~

ABOUT THE AUTHOR

Born long before penicillin and cholesterol were discovered, and defibrillators were invented; before lawyers and government tried to regulate our lives. I drank water from a common tin cup that hung on the school house pump; had no knowledge that lead-base paint was harmful; drove early vehicles that ran on lead gasoline, had mechanical brakes, and no seat belts. Rode bicycles hundreds of miles without a helmet and I thought nothing about hitchhiking a ride with a stranger. Based on scientific facts, anyone my age couldn't or shouldn't have survived except by the knowledge, faith, and trust in our Lord and Savior Jesus Christ.

I was employed as an electric-arc welder for three years during WWII, building LST's (Landing Ship Tanks) at the U. S. Navy Ship Yards in Evansville, Indiana. After the war I attended the Utilities Engineering Institute in Chicago and became a Certified Auto Body Craftsman. Worked in auto body shops and later as Assistance Manager for the Morris 5 & 10 stores in Salem, Indiana and Mt. Vernon, Illinois. Did a stint one summer in door-to-door sales of Hot Point and GE appliances.

From 1950 thru 1987, I was employed in consumer finance for American Security Loans, American Monroe Corp., and Personal Finance Co., a total of 37 years.

My wife, Edna (Satkamp) and I celebrated our 62nd., wedding anniversary June 17, 2006 and was looking forward to our 63rd, when she answered the Lord's call on May 24, 2007. We are the parents of three children, grandparents of seven and great grand parents of six.

My hobbies have been restoring old vehicles and writing short stories, several of which have been published.

Our happiness has been in the art of knowing how to do without what we couldn't afford.

Clendel F. Williams

ORDER FORM

ECHOES OF FREEDOM

TO ORDER ADDITIONAL COPIES PLEASE FILL OUT THE FORM BELOW OR MAKE A COPY AND MAIL TO:

CLENDEL WILLIAMS
5651 N. White River Rd.
Campbellsburg, IN 47108 - 6023
Ph. 1-812-755-4776

For multiple order discount (five or more books), write to the above address for bulk wholesale price.

Please send _______ copies @ $15.00 each, plus
$5.00 mailing fee first copy. $1.00 for additional copies, in the Continental United States.
Alaska, Hawaii, Canada add $6.00; foreign, add $10.00

NAME____

ADDRESS______________________CITY____________________

STATE_________ZIP_____PHONE (OPTIONAL) _______________

SEND A GIFT COPY TO A FRIEND

Autograph to:

NAME __________________ ADDRESS_______________________

CITY ___________________ STATE _________ ZIP______________